# CHESS AND INDIVIDUAL DIFFERENCES

Research from the neurosciences and behavioural sciences highlights the importance of individual differences in explaining human behaviour. Individual differences in core psychological constructs, such as intelligence or personality, account for meaningful variations in a vast range of responses and behaviours. Aspects of chess have been increasingly used in the past to evaluate a myriad of psychological theories, and several of these studies consider individual differences to be key constructs in their respective fields. This book summarizes the research surrounding the psychology of chess from an individual- differences perspective. The findings accumulated from nearly forty years' worth of research about chess and individual differences are brought together to show what is known – and still unknown – about the psychology of chess, with an emphasis on how people differ from one another.

Angel Blanch works in the Department of Psychology at the University of Lleida, Catalonia, Spain. His research focuses on individual differences, intellectual performance and data analyses in behavioural science. Angel also serves as an associate editor at *Personality and Individual Differences*, as well as advising on the editorial board for *Psychological Assessment and Stress and Health.*

# CHESS AND INDIVIDUAL DIFFERENCES

ANGEL BLANCH
*Universitat de Lleida*

Shaftesbury Road, Cambridge CB2 8EA, United Kingdom

One Liberty Plaza, 20th Floor, New York, NY 10006, USA

477 Williamstown Road, Port Melbourne, VIC 3207, Australia

314–321, 3rd Floor, Plot 3, Splendor Forum, Jasola District Centre, New Delhi – 110025, India

103 Penang Road, #05–06/07, Visioncrest Commercial, Singapore 238467

Cambridge University Press is part of Cambridge University Press & Assessment, a department of the University of Cambridge.

We share the University's mission to contribute to society through the pursuit of education, learning and research at the highest international levels of excellence.

www.cambridge.org
Information on this title: www.cambridge.org/9781108469456

DOI: 10.1017/9781108567732



First published 2021
First paperback edition 2022

*A catalogue record for this publication is available from the British Library*

*Library of Congress Cataloging-in-Publication data*
Names: Blanch, Angel, 1967– author.
Title: Chess and individual differences / Angel Blanch, Universitat de Lleida.
Description: New York, NY : Cambridge University Press, 2021. | Includes bibliographical references and index.
Identifiers: LCCN 2020031750 | ISBN 9781108476041 (hardback) | ISBN 9781108567732 (ebook)
Subjects: LCSH: Chess – Psychological aspects. | Chess – Social aspects. | Pattern perception. | Individual differences.
Classification: LCC GV1448 .B53 2021 | DDC 794.1–dc23
LC record available at https://lccn.loc.gov/2020031750

ISBN 978-1-108-47604-1 Hardback
ISBN 978-1-108-46945-6 Paperback

# CONTENTS

# FIGURES

# TABLES

# PREFACE

A considerable body of research within several fields of neurosciences and behavioural sciences has highlighted the crucial importance of individual differences in explaining human behaviour. Individual differences in core psychological constructs such as intelligence or personality account for meaningful variations in a vast diversity of responses and behaviours. Some aspects of the game of chess have been used in the past to evaluate a myriad of psychological theories. Several of these studies consider individual differences as key constructs in their respective fields of research. This book summarizes the latest research about the psychology of chess from an individual differences approach. The volume provides a comprehensive overview of the findings accumulated through nearly forty years of research into chess and individual differences. This volume, *Chess and Individual Differences*, organizes a complete perspective in terms of what is already known and what remains unknown about the psychology of chess, with an emphasis on individual differences.

ACKNOWLEDGEMENTS

Writing this book would have been impossible without the help of the Cambridge staff. In particular, I am gratefully indebted to Janka Romero, Emily Watton, and Jessica Norman for providing assistance throughout the process. Special thanks go to Guillermo Campitelli, who elaborated extensive and priceless feedback on an earlier draft of the manuscript. My greatest debt, however, is to my wife, Loles, and to my two-year-old daughter, Petra, who stoically withstood the time and effort spent on the book and gave me comfort in the meanwhile.

# 1

# Introduction

Several facets of the game of chess have been used in the past to model and evaluate a myriad of psychological theories in a variety of empirical studies. Most of these studies have taken either an experimental or a correlational approach (Table 1.1). Over half a century ago Lee Cronbach examined in detail the evolution of empirical psychology stemming from these two lines of work (Cronbach, 1957). Cronbach contended that a combination of the experimental and correlational approaches would be the most rewarding for advancing psychology, in both basic and applied research. Analogous arguments have repeatedly been brought up, while advocating for a greater degree of cooperation between cognitive scientists and differential psychologists regarding the study of human intelligence (Deary, 2001). Individual differences in several psychological attributes other than intelligence are critical for understanding the behaviour of people. In the past forty years there has been growing interest in the role of these individual differences, because they appear to modulate human behaviour in important domains such as work, health, and education.

Chess can provide a commensurate model of human behaviour, akin to the *Drosophila* model in the biological sciences (Simon & Chase, 1973). Chess has typically been used in terms of the experimental approach to model several theories concerned with cognitive psychology topics. Moreover, the studies carried out in the domain of chess have also increasingly suggested that there are individual differences in several human behavioural attributes, such as brain functioning, memory, thinking, decision-making, intellectual human performance, personality, and motivation. This book compiles and describes this latter body of research.

## 1.1 A Very Brief Opening to the Game of Chess

The origins of the game of chess can be traced back to ancient India around the sixth century AD. Chess travelled first to the West, then, later, to the rest of the world. Nowadays chess has become the universal intellectual game par excellence, practised by millions of individuals of diverse nationalities, ages, and backgrounds. Chess is played on an eight by eight squared board, divided into thirty-two light squares and thirty-two dark squares. Each square is uniquely

Table 1.1 *Overview of the two main approaches to psychological research*

| | Experimental | Correlational |
|---|---|---|
| Aim | Functional analyses of psychological processes | Analysis of individual differences and regularities in behaviour |
| Unit of analysis | Cognitive processes | Psychological traits |
| Hypotheses | Inference | Covariation |
| Research design | Experimental | Ex-post-facto<br>Probabilistic |
| Data analyses | ANOVA<br>ANCOVA<br>MANOVA | Correlation<br>Factor analysis<br>Causal analyses |
| Validity | Internal | External |

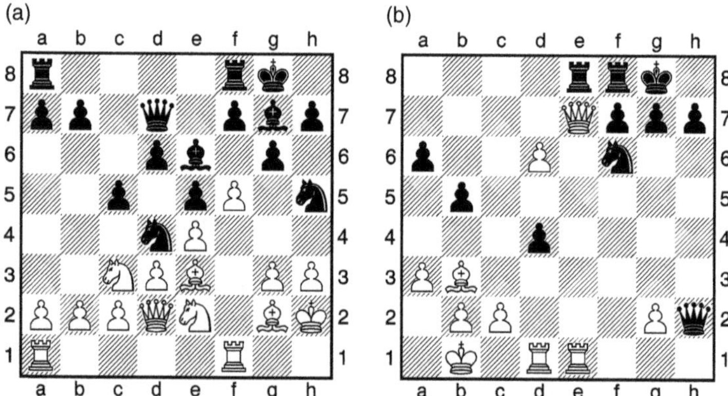


**Figure 1.1** A chess game with all intervening pieces in action (left diagram); a chess problem with white to play and win (right diagram; taken from a game between Velmirovic and Csom in Amsterdam, 1974)

identified by a coordinate system using Latin letters from a to h and numbers from 1 to 8. This board imitates a battlefield on which two armies, one black and one white, confront each other in a merciless fight. Each of the two armies comprises eight pawns, two rooks (R), two knights (N), two bishops (B), a queen (Q), and a king (K). The left diagram in Figure 1.1 shows an ongoing typical chess clash with all these intervening pieces. The specific moves of all pieces are described briefly in the Glossary, together with the value of each piece, indicated by the points usually assigned to it. The aim of the game consists in checkmating the opponent's king. The army that first checkmates the enemy king wins.

The basic rules of the game are very simple and very easy to learn, even at younger ages and at any educational level. Yet the game as a whole becomes extremely complex. There are literally several millions of millions of different combinations among the contending pieces in a single chess game. These combinations can be represented with a chess tree, an informational device in which the solution is the path leading to victory. A chess tree typically generates a massive and unmanageable amount of combinations ($10^{120}$) even for the most powerful and fastest computer chess engines, let alone for human beings (Shannon, 1950). Because each of the pieces involved in the game obeys different movements, the game is intellectually demanding, while requiring the interplay of a variety of major psychological attributes and processes, such as perception, memory, reasoning, decision-making, problem solving, will, motivation, interests, and creativity.

Consider, for instance, the right diagram in Figure 1.1. This represents a typical chess problem with the white forces to play and win. There is an efficient sequence leading to the white victory that comprises five precise moves, with an average time limit to solve it of about ten minutes. The correct sequence of moves in algebraic notation is shown below, for both white and black pieces:

| | **White** | **Black** |
|---|---|---|
| 1 | B×f7+ | R×f7 |
| 2 | Q×e8+ | N×e8 |
| 3 | R×e8+ | Rf8 |
| 4 | d7 | Qd6 |
| 5 | Rf1!! | 1–0 |

Each of the five chess moves comprise two plies: one ply for white, and one ply for black. The ply corresponding to black in the fifth move indicates that white has won the game (scoring one point), however, because, after the last ply of white (Rf1), there is no possible legal move by black to avoid being checkmated at the very next move by white. Capital letters stand for the specific chess piece being moved and the × symbol indicates that a piece captures an opponent's piece. For instance, the first ply for white (B×f7+) indicates that the white bishop (B), initially placed in the b3 square, is capturing the pawn located in the f7 square. The + symbol indicates that the black king is placed in check. A ply depicting a single square only indicates a pawn move. For instance, the fourth ply for white (d7) indicates that the pawn placed in the square d6 advances to the square d7. The double exclamation mark in the fifth ply for white (Rf1!!) indicates a brilliant and very strong move. In this specific game, it was a *coup de grâce* move, winning the game.

People may differ greatly in terms of their chances of finding out this sequence of moves. If you are a proficient chess player at the master level, you may be able to 'see' the sequence at a glance. It could also be the case that you may remember this position because you have already studied it in the past, during your long chess career. On the other hand, if you are a typical club chess player with a moderate level of chess skill, you might invest the suggested amount of time, but you may end up unable to figure out what the correct solution is at all. If you are a beginner chess player, or you just know the basic chess rules, the likelihood of experiencing serious difficulties in finding the solution may be so great as to be insurmountable. This is a very basic example of individual differences in chess performance and chess skill.

## 1.2 Overview of This Book

Nowadays there is a considerable volume of chess studies that have highlighted noteworthy individual differences. For example, some of these chess studies use problems such as that shown in Figure 1.1 as experimental stimuli. This book is an attempt to compile and summarize the latest research about the psychology of chess with a focus on individual differences. Besides, this volume aims to provide an overview of the findings from more than forty years of research, from the mid-1970s to date, about chess and individual differences. This body of research has sometimes yielded inconclusive and even controversial results, suggesting, for instance, that the development of chess skill over time may largely depend on the combination of individual differences in several traits or broad clusters of traits. This book organizes the body of knowledge that uses chess as a model environment, while providing useful scientific information about a variety of individual differences in brain functioning, intelligence, personality, expertise, and sex, and in applied fields such as business, health, and education.

The book is mainly aimed at scholars within the broad spectrum of the social and behavioural sciences who have an interest in the psychology of chess. The book can be of interest to psychologists, sociologists, educators, neuroscientists, and behavioural scientists in general. The chapters are intended to cover the topics typically addressed by social scientists interested in individual differences working in a diversity of fields. Those researchers and academics working in brain functioning, human abilities, and personality may find the book appealing. Moreover, the book may also arouse the curiosity of researchers and academics working with topics such as expertise, sex differences, and education, or with a focus on applied fields. In addition, the book may also be of interest for people who play chess themselves. In particular, chess players wishing to gain a more in-depth understanding of the scientific work undertaken with chess as a model domain from a psychological approach may find some stimulating information within these pages.

Chapter 2 describes the Elo chess rating. What makes chess an optimum field for the study of individual differences is the availability of this objective quantitative measure to gauge a player's chess strength. The Elo rating system is by far the most popular and accepted indicator worldwide for quantifying accurately individual differences in chess skill. Every chess player participating regularly in rated chess tournaments holds an Elo rating. The Elo rating changes according to the outcomes of the games played within a given time period, while considering the Elo rating of the opponents. The chapter describes how the Elo ratings of thousands of chess players are kept and periodically updated. It also outlines the updating mechanisms and some basic statistics of the Elo rating. In addition, the chapter describes some recent alternatives to rating chess skill, such as the Universal Rating System (URS). Appendix 1 summarizes the studies that have used the Elo rating as related to a variety of human behaviours.

Chapters 3 and 4 provide an overview of the main findings from the cognitive and the individual differences approach to the psychology of chess, respectively. Chapter 3 reviews the main research findings from the cognitive or experimental paradigm within psychology, which originated with the precursor scientific works about the psychology of chess. Three main basic facets of human behaviour have been addressed within this general approach: perception, memory, and thinking. The main conclusions from this extensive body of research can be summarized by emphasizing the role of individual differences. Chapter 4 outlines the main tenets and constructs of differential psychology, the discipline that studies individual differences in behaviour relevant for central social realms such as health, education, and work. The chapter is structured around three main themes. First, it describes the characterization and appraisal of individual differences. Second, the PPIK theory is suggested as an optimal starting point to conceptualize and examine individual differences. This framework comprises traits from four broad dimensions: intelligence as process, personality, interests, and intelligence as knowledge. Third, the chapter closes by addressing the old but compelling debate about the heredity versus environment dichotomy in explaining complex human intellectual behaviour.

Chapter 5 describes the studies addressing human biological factors in chess, with a focus on psychophysiology and brain imaging. Human psychophysiology is a multi-faceted and complex phenomenon. The game of chess has provided a proper domain for the study of the central psychophysiological mechanisms underlying psychological processes such as stress, emotion evaluation, and decision-making. Moreover, novel technologies designed to provide high-resolution brain imaging are being increasingly used to explain human behaviour. These technologies have also been used with chess players to examine the interrelationships of brain and cognitive functioning, and with personality and intelligence factors. In particular, this chapter outlines the research undertaken with electroencephalography (EEG), functional magnetic resonance imaging (fMRI), and positron emission tomography (PET). The chapter summarizes this body of evidence while underlining the most significant conclusions that may be derived from this intriguing and thought-provoking field of research.

Chapter 6 provides an account of the studies addressing chess and intelligence. Human intelligence is one of the main general objects of study in individual differences research. There are indeed multiple models about and approaches to human intelligence, which are briefly described within this chapter. Chess has been typically associated with a high level of intelligence. Whether chess players are more intelligent on average than the general population is a recurrent question that has elicited a considerable body of research. There are unsettled issues as to what constitute the most advantageous cognitive abilities required in chess, and whether playing chess makes people smarter. These topics have been addressed with both children and adults. The scientific evidence in connection with this topic is inconclusive, however, and controversial in some instances. This chapter addresses these matters of contention by summarizing the state of the art in this particularly cogent field of research. The final section in the chapter includes novel empirical findings comparing chess skill and chess motivation in the prediction of chess performance, suggesting that non-cognitive traits might also be influential for chess performance.

Chapter 7 analyses what is already known about chess and human personality. Personality is the other main broad domain addressed within the general framework of individual differences. In contrast with intelligence, however, the body of research concerning the personality of chess players is rather scarce. There have been some interesting findings recently, however, and these are summarized within this chapter. After describing briefly the main approaches to addressing human personality, some questions addressed in this chapter are whether personality influences chess playing style, or whether a chess player's personality differs in some special way from that of other people. In addition, whether personality factors may interact with cognitive abilities in chess players is an interesting and relatively novel topic. The chapter closes by presenting novel data about the interplay between personality, motivation, and emotional regulation in predicting chess skill.

Chapter 8 analyses expertise, one of the most prolific fields in empirical research using chess as a model domain. Expertise is of great importance in several realms of human intellectual activity. The role of practice in the development of chess expertise is reviewed in detail in this chapter. Moreover, the role of practice is contrasted with talent, because the deliberate practice approach has advanced the idea that expert performance depends exclusively on practice. A consistent body of evidence suggests that deliberate practice alone is unable to explain the individual variability in chess expertise, however. The present chapter addresses this controversy by framing these findings in the nature versus nurture debate, one of the central themes within individual differences research. Furthermore, this chapter also explores age-related cognitive decline in human intellectual activity, which appears to occur to a lesser extent in the chess domain. For instance, recent findings suggest in particular two interrelated factors that may be highly relevant in preventing cognitive decline in chess: the level of expertise attained, and the amount of tournament activity.

Chapter 9 tackles the issue of sex differences in chess. On average, men tend to start earlier, perform at a higher level, and persist longer than women in the chess domain. Moreover, women are highly underrepresented in chess, which is also apparent in several other domains, such as those connected with STEM fields (science, technology, engineering, and mathematics). The marked difference in the number of men and women participating in chess has led to the assumption that the differences in chess performance between men and women are attributable to a statistical effect derived from the differences in participation rates. In contrast, other findings suggest that men might have an innate advantage in terms of chess playing, enhanced by certain cultural factors. These two points of view are addressed in this chapter. The alternative explanation to the marked disparity in chess participation and performance between the two sexes may be related to the participation of men and women in STEM fields. In addition, there are some noticeable differences in the chess playing of men and women, even though women are able to play very strong chess, just like men. The chapter closes by presenting a statistical analysis with data from the chess domain, which relates to sex differences in performance at different levels of practice. The findings from this analysis suggest that sex differences in the Elo ratings tend to increase with increasing practice, pointing to factors other than practice as the underlying causes of these sex differences.

Chapter 10 deals with the applications of chess in three major fields of human activity: business, health, and education. Chess has been used in the business field with two main aims. First, chess has been used for educational purposes to teach and consolidate concepts connected with this discipline. Second, some studies have used chess as a model to evaluate game-theory aspects of the game. The game of chess has also been increasingly used to address health-related problems such as attention deficit hyperactivity disorder (ADHD), neurodegenerative disorders, and schizophrenia. Moreover, chess has become an increasingly popular pedagogical method in several school settings across the world. A number of studies claim that chess training entails several educational benefits for core academic subjects such as languages and mathematics, and also for concentration and self-control, or the development of socio-affective competences. Several of the instructional experiences that use chess to enhance these behaviours are described in this chapter. Some recent studies suggest that significantly higher levels of academic performance for schoolchildren and adolescents are associated with chess-based teaching or the practice of chess on a regular basis, when compared with those students who are not involved in chess playing or chess instruction. Another set of studies have questioned the purported benefits of chess training for formal education, however. From this latter point of view, there are both conceptual and methodological concerns that compromise to a great extent the available evidence about the association of chess training with academic achievement. Two of these issues relate to the transfer of abilities across domains, and to the concept of statistical power.

Chapter 11 is the closing chapter of this book. This chapter argues why chess has become an interesting domain to address topics of interest for individual differences research. It also summarizes the most robust available evidence to date by outlining the key findings, while suggesting some tentative and potentially promising steps for advancing the field.

# 2

# Quantifying Chess Skill

What makes chess an optimum field for the study of individual differences is the availability of an objective quantitative measure of a player's chess strength. This is an important asset compared with other applied domains, because they lack such a systematic indicator of skill. Although several indicators quantify accurately chess skill, the Elo rating system is the more popular chess skill indicator, accepted worldwide. For example, the Elo rating system is useful in the organization of formal chess tournaments, such as in pairing players of equivalent chess strength, or in restricting participation in chess tournaments to a given chess strength level or to groupings of players with different levels of chess skill. Because the Elo rating is an interval scale, it lacks a true zero, though it allows the quantification of an objective difference between each value.

Every chess player participating regularly in rated tournaments holds an Elo rating. The Elo rating is a dynamic indicator that depends on the outcomes of the games played within a given time period, taking into account the Elo rating of the opponents. Such a system has been deemed highly appropriate to track changes in the variability of its scale values, which might be useful for addressing an extensive variety of research problems within differential psychology or individual differences research (Batchelder & Bershad, 1979; Howard, 2006).

A sense of the variability in chess skill as measured by the Elo rating can be gleaned by looking at the world maps displayed within Appendix 2. These maps represent data for 118 countries obtained from the December 2018 list of the World Chess Federation (Fédération Internationale des Échecs, FIDE). The first map shows the mean Elo rating, and the second map shows the number of chess grandmasters by country. There are only three countries with a mean Elo rating above 2700 Elo points – Russia, China, and the United States – and twelve countries with a mean Elo rating above 2600 Elo points: Azerbaijan, Ukraine, India, France, Armenia, Hungary, the Netherlands, Poland, the United Kingdom, Israel, Germany, and Spain. Cross-country differences are more pronounced, however, when looking at the number of grandmasters in the second map. Here, Russia holds a noteworthy advantage over the rest of the countries, with 251 grandmasters, in front of the United States, with ninety-eight, Germany, with ninety-six, and Ukraine, with ninety-one. There is then a group of five countries with between fifty and fifty-seven grandmasters: Serbia, Hungary, India, Spain, and France. In contrast, the world regions with the lower mean Elo ratings and number of grandmasters correspond to Africa, and several countries in Central and South America, and Asia. The two maps evidence the universality of the game, which is surely unparalleled by any other game of its kind.

## 2.1 Elo Rating Lists

The systematic updated records of the Elo ratings of chess players from all over the world allow the study of individual differences in intellectual performance from an objective point of view. Every chess player participating regularly in rated chess tournaments of any kind holds an Elo rating that ranges from approximately 1,200 to about 2,850 points, with higher scores being indicative of a higher level of chess strength (Elo, 1978; Glickman, 1995; Glickman & Chabris, 1996; Glickman & Jones, 1999). Chess federations worldwide keep and update periodic records of the Elo ratings of their respective players. In addition, players participating in international tournaments hold the Elo rating of the respective player's country or local chess federation, and the international Elo rating assigned by the World Chess Federation (Fédération Internationale des Échecs: FIDE). Elo ratings from different chess federations tend to be highly correlated. There are even Elo rating lists from a variety of computer chess engines. Figure 2.1 shows part of the Elo rating lists of the FIDE, the Spanish Chess Federation, and the Catalan Chess Federation, and the Elo ratings of 353 computer chess engines.

The lists from the World and Catalan Chess Federations and from computer engines are ordered by the rank of the strongest players. The World Chess Federation list shows the ten strongest players. The Catalan Chess Federation list indicates the chess title and sex of each player (GM: Grandmaster; M: Male). The computer list shows the number of games played and the percentage of winning outcomes. The Spanish Chess Federation list is in alphabetical order by the player's surname, and it also includes the number of games played in the given period, the year of birth, the title, whether the player is active or inactive (A, I), and the previous Elo rating. For instance, the current Elo of the first player in this list is 1851 points, while his previous Elo was 1842. Therefore, the player has gained nine Elo points in this latter Elo update. In contrast, the player with Id. FEDA #26 has a current Elo of 2177, while his previous Elo was 2181. Therefore, this player has lost four Elo points in this latter Elo update.

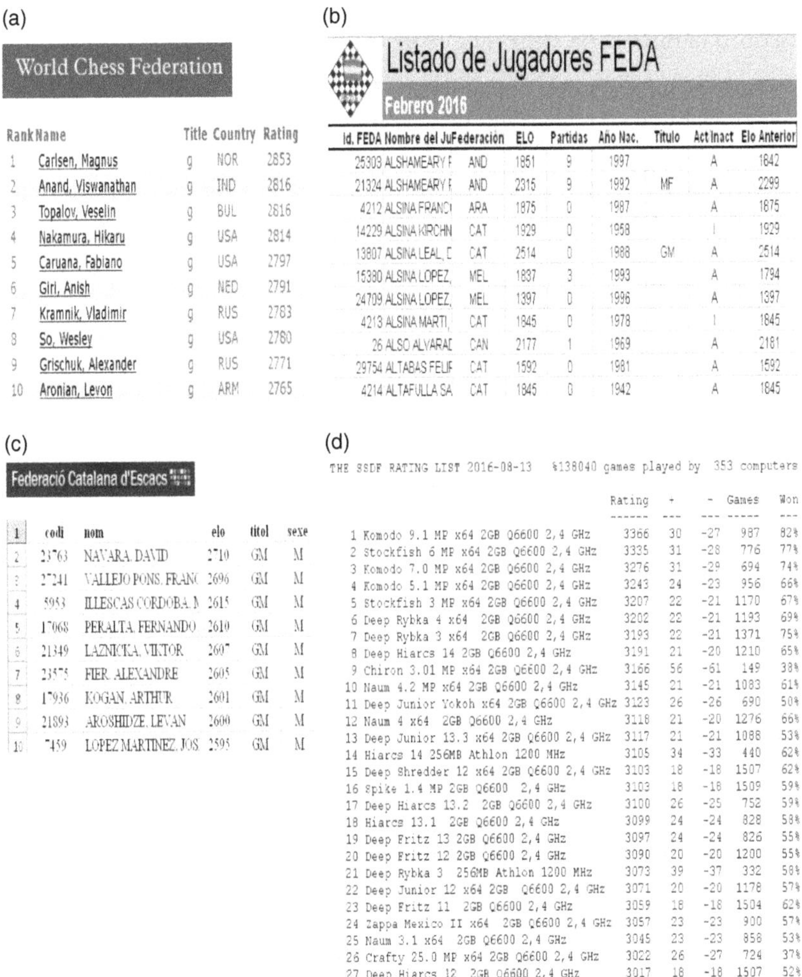

(a)

World Chess Federation

| Rank | Name | Title | Country | Rating |
|---|---|---|---|---|
| 1 | Carlsen, Magnus | g | NOR | 2853 |
| 2 | Anand, Viswanathan | g | IND | 2816 |
| 3 | Topalov, Veselin | g | BUL | 2816 |
| 4 | Nakamura, Hikaru | g | USA | 2814 |
| 5 | Caruana, Fabiano | g | USA | 2797 |
| 6 | Giri, Anish | g | NED | 2791 |
| 7 | Kramnik, Vladimir | g | RUS | 2783 |
| 8 | So, Wesley | g | USA | 2780 |
| 9 | Grischuk, Alexander | g | RUS | 2771 |
| 10 | Aronian, Levon | g | ARM | 2765 |

(b)

Listado de Jugadores FEDA

Febrero 2016

| Id. FEDA | Nombre del Ju | Federación | ELO | Partidas | Año Nac. | Título | Act Inact | Elo Anterior |
|---|---|---|---|---|---|---|---|---|
| 25303 | ALSHAMEARY F | AND | 1851 | 9 | 1997 | | A | 1842 |
| 21324 | ALSHAMEARY F | AND | 2315 | 9 | 1992 | MF | A | 2299 |
| 4212 | ALSINA FRANC | ARA | 1875 | 0 | 1987 | | A | 1875 |
| 14229 | ALSINA KIRCHN | CAT | 1929 | 0 | 1958 | | I | 1929 |
| 13807 | ALSINA LEAL, D | CAT | 2514 | 0 | 1988 | GM | A | 2514 |
| 15380 | ALSINA LOPEZ, | MEL | 1837 | 3 | 1993 | | A | 1794 |
| 24709 | ALSINA LOPEZ, | MEL | 1397 | 0 | 1996 | | A | 1397 |
| 4213 | ALSINA MARTI, | CAT | 1845 | 0 | 1978 | | I | 1845 |
| 26 | ALSO ALVARAD | CAN | 2177 | 1 | 1969 | | A | 2181 |
| 29754 | ALTABAS FELIP | CAT | 1592 | 0 | 1981 | | A | 1592 |
| 4214 | ALTAFULLA SA | CAT | 1845 | 0 | 1942 | | A | 1845 |

(c)

Federació Catalana d'Escacs

| 1 | codi | nom | elo | titol | sexe |
|---|---|---|---|---|---|
| 2 | 23763 | NAVARA, DAVID | 2710 | GM | M |
| 3 | 27241 | VALLEJO PONS, FRANC | 2696 | GM | M |
| 4 | 5953 | ILLESCAS CORDOBA, M | 2615 | GM | M |
| 5 | 17068 | PERALTA, FERNANDO | 2610 | GM | M |
| 6 | 21349 | LAZNICKA, VIKTOR | 2607 | GM | M |
| 7 | 23575 | FIER, ALEXANDRE | 2605 | GM | M |
| 8 | 17936 | KOGAN, ARTHUR | 2601 | GM | M |
| 9 | 21893 | AROSHIDZE, LEVAN | 2600 | GM | M |
| 10 | 7459 | LOPEZ MARTINEZ, JOS | 2595 | GM | M |

(d)

THE SSDF RATING LIST 2016-08-13 %138040 games played by 353 computers

| | | Rating | + | - | Games | Won |
|---|---|---|---|---|---|---|
| 1 | Komodo 9.1 MP x64 2GB Q6600 2,4 GHz | 3366 | 30 | -27 | 987 | 82% |
| 2 | Stockfish 6 MP x64 2GB Q6600 2,4 GHz | 3335 | 31 | -28 | 776 | 77% |
| 3 | Komodo 7.0 MP x64 2GB Q6600 2,4 GHz | 3276 | 31 | -29 | 694 | 74% |
| 4 | Komodo 5.1 MP x64 2GB Q6600 2,4 GHz | 3243 | 24 | -23 | 956 | 66% |
| 5 | Stockfish 3 MP x64 2GB Q6600 2,4 GHz | 3207 | 22 | -21 | 1170 | 67% |
| 6 | Deep Rybka 4 x64 2GB Q6600 2,4 GHz | 3202 | 22 | -21 | 1193 | 69% |
| 7 | Deep Rybka 3 x64 2GB Q6600 2,4 GHz | 3193 | 22 | -21 | 1371 | 75% |
| 8 | Deep Hiarcs 14 2GB Q6600 2,4 GHz | 3191 | 21 | -20 | 1210 | 65% |
| 9 | Chiron 3.01 MP x64 2GB Q6600 2,4 GHz | 3166 | 56 | -61 | 149 | 38% |
| 10 | Naum 4.2 MP x64 2GB Q6600 2,4 GHz | 3145 | 21 | -21 | 1083 | 61% |
| 11 | Deep Junior Yokoh x64 2GB Q6600 2,4 GHz | 3123 | 26 | -26 | 690 | 50% |
| 12 | Naum 4 x64 2GB Q6600 2,4 GHz | 3118 | 21 | -20 | 1276 | 66% |
| 13 | Deep Junior 13.3 x64 2GB Q6600 2,4 GHz | 3117 | 21 | -21 | 1088 | 53% |
| 14 | Hiarcs 14 256MB Athlon 1200 MHz | 3105 | 34 | -33 | 440 | 62% |
| 15 | Deep Shredder 12 x64 2GB Q6600 2,4 GHz | 3103 | 18 | -18 | 1507 | 62% |
| 16 | Spike 1.4 MP 2GB Q6600 2,4 GHz | 3103 | 18 | -18 | 1509 | 59% |
| 17 | Deep Hiarcs 13.2 2GB Q6600 2,4 GHz | 3100 | 26 | -25 | 752 | 59% |
| 18 | Hiarcs 13.1 2GB Q6600 2,4 GHz | 3099 | 24 | -24 | 828 | 58% |
| 19 | Deep Fritz 13 2GB Q6600 2,4 GHz | 3097 | 24 | -24 | 826 | 55% |
| 20 | Deep Fritz 12 2GB Q6600 2,4 GHz | 3090 | 20 | -20 | 1200 | 55% |
| 21 | Deep Rybka 3 256MB Athlon 1200 MHz | 3073 | 39 | -37 | 332 | 58% |
| 22 | Deep Junior 12 x64 2GB Q6600 2,4 GHz | 3071 | 20 | -20 | 1178 | 57% |
| 23 | Deep Fritz 11 2GB Q6600 2,4 GHz | 3059 | 18 | -18 | 1504 | 62% |
| 24 | Zappa Mexico II x64 2GB Q6600 2,4 GHz | 3057 | 23 | -23 | 900 | 57% |
| 25 | Naum 3.1 x64 2GB Q6600 2,4 GHz | 3045 | 23 | -23 | 858 | 53% |
| 26 | Crafty 25.0 MP x64 2GB Q6600 2,4 GHz | 3022 | 26 | -27 | 724 | 37% |
| 27 | Deep Hiarcs 12 2GB Q6600 2,4 GHz | 3017 | 18 | -18 | 1507 | 52% |
| 28 | Deep Shredder 11 x64 2GB Q6600 2,4 GHz | 3013 | 21 | -21 | 1088 | 50% |

**Figure 2.1** Elo rating lists of the World (a), Spanish (b), and Catalan Chess Federations (c), and Elo rating list of top twenty-eight computer chess engines out of a list of 353 engines (d)

## 2.2 Updating Mechanism and Basic Statistics of the Elo Rating

The Elo rating is a dynamic indicator that depends on the outcomes of the games played within a given period considering also the Elo ratings of the opponents. To illustrate, the elements implied in the calculation and updating of the Elo rating after the outcome of a chess game between two players can be described in three main steps (Elo, 1978; Glickman, 1995):

1. There are three possible outcomes arising from a chess game: win = 1, draw = 0.5, defeat = 0.
2. The expected score (*E*) in a chess game between players *A* and *B* with ratings $R_A$ and $R_B$ can be calculated for player *A* with the expression in Equation 2.1:

$$E = \frac{10^{R_A/400}}{10^{R_A/400} + 10^{R_B/400}} \tag{1}$$

3. The update of the Elo rating is calculated with the expression in Equation 2.2. This includes a previous Elo rating ($r_{pre}$), the Elo rating after a chess tournament ($r_{post}$), a constant value (*K*), the sum of points obtained in the tournament (*S*), and the sum of expected scores in each game ($S_{exp}$):

$$r_{post} = r_{pre} + K(S - S_{exp}) \tag{2}$$

In a single chess game, a win scores one point, a defeat scores zero points, and a draw scores half a point. The expression in Equation 2.1 can be conceived as the actual probability of winning a chess game considering the Elo rating of both opponents. The Elo rating is indeed an accurate predictor of the outcome of a chess game. A stronger player in terms of a higher Elo rating has increased chances of scoring one point when playing against a weaker player. In contrast, a weaker player in terms of a lower Elo rating sees his or her chances of scoring one point greatly decreased when playing against a stronger player. The expression in Equation 2.2 serves to update the Elo rating in accordance with the performance of a player within a given period. The new and updated Elo rating ($r_{post}$) is calculated by summing the observed previous Elo rating ($r_{pre}$) and the term *K*(*S* – *Sexp*). The value of *K* corresponds to an attenuation factor that represents the amount of weight allotted to a new Elo rating given an old Elo rating – that is, the maximum number of points that increase or decrease the rating from the outcome of a single chess game. Larger *K* values allow greater changes in Elo ratings. Usually, younger and less experienced players tend to have higher attenuation *K* values than older and more experienced players. The value (*S* – *Sexp*) indicates the discrepancy observed between the actual points (*S*) obtained within a given period and the expected points (*Sexp*), calculated with the expression in Equation 2.1 in accordance with the Elo ratings of the corresponding opponents in the chess games played within this period. Positive values in (*S* – *Sexp*) indicate that the player performed above what was expected, whereas negative values indicate that the player

Table 2.1 *Example of the Elo rating updating in one chess game between players KS versus AB, and JP versus LQ*

| Player | Elo rating | *K* | Expected score | Actual score | Game outcome | Elo update |
|---|---|---|---|---|---|---|
| KS | 2544 | 10 | 0.97 | 1 | Win | 2544 |
| AB | 1936 | 15 | 0.03 | 0 | Defeat | 1935 |
| KS | 2544 | 10 | 0.97 | 0 | Defeat | 2534 |
| AB | 1936 | 15 | 0.03 | 1 | Win | 1951 |
| KS | 2544 | 10 | 0.97 | 0.5 | Draw | 2539 |
| AB | 1936 | 15 | 0.03 | 0.5 | Draw | 1943 |
| JP | 2064 | 15 | 0.44 | 1 | Win | 2071 |
| LQ | 2106 | 15 | 0.56 | 0 | Defeat | 2098 |
| JP | 2064 | 15 | 0.44 | 0 | Defeat | 2056 |
| LQ | 2106 | 15 | 0.56 | 1 | Win | 2113 |
| JP | 2064 | 15 | 0.44 | 0.5 | Draw | 2065 |
| LQ | 2106 | 15 | 0.56 | 0.5 | Draw | 2105 |

performed below what was expected. Therefore, the updated Elo rating will increase or decrease accordingly.

Table 2.1 shows two examples of the Elo rating update in a hypothetical chess game. In the first example, concerning players KS and AB, there is a considerable difference between the Elo ratings of the two players, 608 points, while player KS holds a lower K value of ten compared with the K value of fifteen for player AB. The first two rows in the table show the most likely situation after the game: a victory of the player with the higher Elo rating. The Elo rating update for this player would be unmodified from the previous Elo rating (Elo update = 2544). On the other hand, the Elo rating for player AB would decrease the previous Elo rating (Elo update = 1935) by just one point. In contrast, with an unlikely defeat of the stronger player, KS, the previous Elo rating decreases by ten points (Elo update = 2534), while for the weaker player, AB, there is an increment of fifteen points (Elo update = 1951). The result of a draw, with 0.5 points for each player, is also more advantageous for the weaker player; it decreases the Elo rating of KS by five points (Elo update = 2539), while increasing the Elo rating of AB by seven points (Elo update = 1943). In the second example, the difference between players JP and LQ is only forty-two Elo points, which is markedly lower than in the previous example, with both players holding a *K* value of fifteen. The three possible outcomes of the game, win, draw, and defeat, indicate a more balanced outcome for each player concerning their corresponding Elo updates. This is perhaps better seen

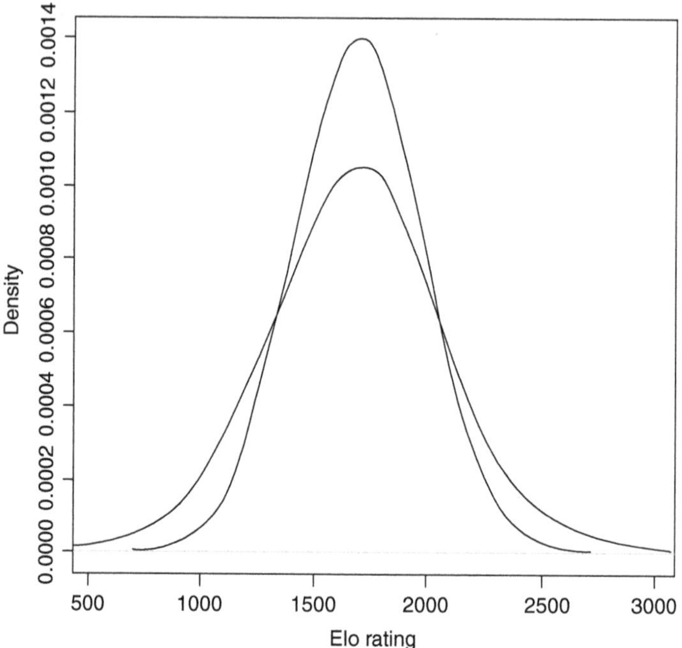


**Figure 2.2** Density plot of the Elo rating with normal (continuous line) and logistic (dotted line) distributions

in the last two rows, which describe the draw situation, which modifies the Elo update by just one point for each player.

The Elo rating system can be framed within the Bradley–Terry model for predicting the comparison of a pair of objects or individuals (Bradley & Terry, 1952), albeit, in its development, the Elo rating system assumed a normal distribution. Nevertheless, the Bradley–Terry model relies instead on logistic distribution assumptions, which is the approach taken by the World and the US Chess Federations (Glickman, 1995). Figure 2.2 shows both normal and logistic probability distributions for a simulated sample of 1,000 players with a mean of 1700 Elo points (*Sd* = 200). There are higher density estimates at the centre of the normal distribution, and longer extended tails for the logistic distribution. On the other hand, Figure 2.3 shows the density plots from four different chess tournaments in four variables: the age of the participants, the number of games prior to the tournament, the tournament outcome, and the Elo rating. The data corresponding to the four tournaments indicate a relatively consistent overlap apart from for the Elo rating, suggesting that the Elo rating was more variable than age, number of games, or tournament

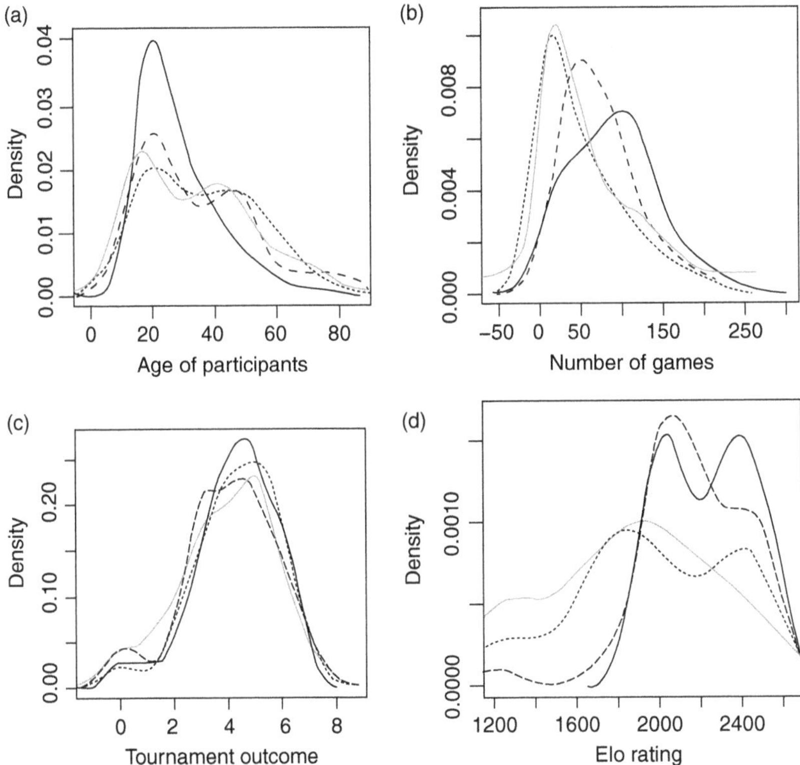


**Figure 2.3** Density plots for the distribution of the Elo ratings in the age of participants (a), number of games (b), tournament outcome (c), and Elo rating (d) of four chess tournaments

outcomes. Table 2.2 displays the descriptive statistics in the Elo rating across the four chess tournaments. The Shapiro–Wilk normality tests for the Elo rating in the four tournaments suggest that the null hypothesis stating that the sample comes from a population with a normal distribution is unsupported by these data.

## 2.3 Alternatives to the Elo Rating of Chess Players

The Elo rating system has gained considerable acceptance and has been used worldwide for over forty years since its inception (Elo, 1978). There are several criticisms and suggestions for improvement, however. For example, it has been recommended that chess ratings should deal with unrated and recently rated players, and that the $K$ attenuation factor should be related to the number of

Table 2.2 *Descriptive statistics of the Elo ratings in four chess tournaments*

| Statistic | Tournament 1 | Tournament 2 | Tournament 3 | Tournament 4 |
|---|---|---|---|---|
| *n* | 107 | 100 | 85 | 81 |
| Min | 1873 | 1200 | 1200 | 1200 |
| Q1 | 2034 | 1591 | 1724 | 2025 |
| Q2 | 2234 | 1878 | 1982 | 2131 |
| Q3 | 2411 | 2143 | 2349 | 2394 |
| Max | 2625 | 2574 | 2596 | 2577 |
| *M* | 2234 | 1847 | 1975 | 2165 |
| *Sd* | 213 | 386 | 393 | 262 |
| *Cv* | 0.10 | 0.21 | 0.20 | 0.12 |
| Skewness | 0.01 | -0.15 | -0.32 | -1.03 |
| Kurtosis | -1.27 | -0.80 | -0.73 | 2.86 |
| *W* | 0.95*** | 0.96** | 0.95** | 0.91*** |

*Notes:* $n$ = sample size; Min: minimum; Q1 to Q3: quartiles 1 to 3; Max: maximum; *M*: mean; *Sd*: standard deviation; *Cv*: coefficient of variation (*Sd/M*); *W*: Shapiro–Wilk normality test ($H_0$: the variable follows a normal distribution); $^{**}p < 0.01$; $^{***}p < 0.001$.

games played (Fenner, Levene, & Loizou, 2012). There are, in addition, other alternatives to rate chess skill by relying on Bayesian methods. For instance, there is a proposal to estimate chess skill from the moves played in an assortment of chess games, rather than from competitive chess outcomes (Di Fatta, McHaworth, & Regan, 2009). Other work has focused on the modelling of draws and the inference of chess skill from chess team outcomes (Herbrich & Graepel, 2006).

Furthermore, there are other approaches raising substantial modifications to the Elo rating system. For example, Chessmetrics is a comprehensive internet database about the rating of the chess skill of chess masters throughout history. Jeff Sonas, the developer and chief engineer of Chessmetrics, proposed an alternative system to calibrate chess skill with alternative methods to the Elo rating (Sonas, 2002). There were four main suggestions regarding the Sonas system: using a more dynamic *K*-factor; dismissing the Elo table and opting for a simpler linear model; including faster time controls, albeit assigning them a lower importance than slower time controls; and calculating the chess ratings on a monthly basis. Moreover, other alternative chess rating systems include the Glicko (Glickman, 1999), the Glicko-2 (Glickman, 2001), and the Universal Rating System (URS), all of them developed by Professor Mark Glickman, a statistician at Harvard University, together with other researchers.

The URS is a particularly appealing system, because it provides a single rating for each individual obtained from the outcomes of games played at different time controls. In the FIDE Elo rating system, chess players can have up to three different Elo ratings stored in three different lists: standard, rapid, and blitz ratings. The three ratings are highly correlated ($r > 0.90$). The standard Elo rating is estimated from the outcomes obtained in standard slow games, which allow at least two hours for each player for their first sixty moves. The rapid Elo rating is estimated from the outcomes obtained in fast games that last more than ten minutes but less than sixty minutes for each player. The blitz Elo rating is estimated from the outcomes obtained in very fast games, which last ten minutes or less for each player. Faster blitz games, therefore, impose serious constraints on the available thinking time, whereas slower standard games allow for a larger amount of thinking time. This fact has been capitalized on for analysing chess thinking strategies such as pattern recognition and search A lower amount of available thinking time is likely to foster a more intensive application of fast pattern recognition, whereas a higher amount of available thinking time is likely to foster a more intensive application of slow search thinking (Burns, 2004; van Harreveld, Wagenmaapplication of slow search thinking (Burns, 2004; van Harreveld, Wagenmakers, & van der Maas, 2007). The URS contemplates the variability of the outcomes in chess games from five minutes (blitz) to two hours (standard), by calculating the degradation of playing skill from slower to faster games. The degradation from slower to faster games is indicated with rapid or blitz gaps with respect to slower games, with higher values being indicative of a higher degradation. Even though they use information from the outcomes in games at different time controls, the designers of the system acknowledge the lower informative value of faster games compared with slower games about the true underlying individual chess skill. Thus, fast game outcomes contribute to the URS to a lesser extent than slow game outcomes. A detailed description of the URS is available through the internet at http://universalrating.com.

## 2.4 Overview of Studies Using the Elo Rating

Because the aforementioned rating systems have only recently been implemented, most of the studies in the psychology of chess have addressed individual differences in chess performance by considering the Elo rating as a central variable of interest. The Elo rating is an appealing indicator that has attracted considerable attention from diverse fields of research. Elo rating data has been used to model animal and macroeconomic behaviour (Albers & de Vries, 2001; Burns, 2004), computerized adaptive testing (Moul & Nye, 2009), and evolutionary algorithms within computing science (Antal, 2013).

For instance, a statistical study has addressed three interesting hypotheses (Breznik & Batagelj, 2011): (1) whether the best players in the world tended to play exclusively among themselves; (2) whether the amount of games played depended on geographical proximity between countries; and (3) whether the advantage of conducting the white pieces should be considered in determining the strength of a chess player. The analyses were based on an extensive sample of chess players ($n$ = 92,731) with one of the FIDE Elo rating lists. The findings in the study support the view that the best players were quite selective concerning chess tournaments. Besides, chess players from different countries played each other more often when they came from neighbouring countries. Finally, conducting the white pieces was judged as an important advantage at higher levels of chess skill, suggesting that it should be also contemplated in the calculation of chess ratings.

The Elo chess rating is a quantitative measure that allows the estimation of individual differences in chess skill accurately and reliably, and it has been extensively used in the research about individual differences in the chess domain. Appendix 1 shows a representative overview of the studies about human behaviour that have used the Elo chess rating or analogous ratings from diverse chess federations, and the factors potentially related to these measures of chess skill ($n$ = 134). For each study there is the journal in which it was published, the number of participants or sample sizes ($N$), and the main conclusions. In some studies, the Elo ratings were used to split a given sample of chess players into diverse levels of expertise, and, subsequently, to apply different batteries of psychological tests or experimental tasks. Other studies analysed the association of the Elo rating with a variety of outcomes, such as brain and psychophysiological activity, the quality of chess playing, perception, memory, reasoning, verbal and processing speed performance, intelligence, personality, sex and age differences, practice, expertise, and learning. In what remains of the present book, each of these themes will be addressed in depth, by delving into the main correlates of the Elo chess rating that have been found in this body of research.

# 3

# Cognition

The cognitive approach to the psychology of chess encapsulates that branch of research interested in the mental processes elicited when playing chess, or when working with chess material within an experimental setting. Empirical studies conducted using this approach generally enquire about the mental processes implied in the intellectual activity of chess that are common to all individuals. That these processes might largely depend on individual differences in chess skill was readily envisaged in earlier studies within the field, however (Binet, 1893, 1894; Cleveland, 1907; de Groot, 1965; Djakow, Petrowski, & Rudik, 1927).

These early works agreed fairly well about the main psychological factors implied in chess playing. For example, the foundational work by Alfred Binet addressed mental processes elicited during simultaneous blindfold chess playing, a stringent chess modality whereby one person plays several opponents without actually visualizing the boards (Binet, 1894). The performance in blindfold chess playing should be likely to depend on individual differences in knowledge and experience in the domain of chess, together with imagination, and memory. Other central psychological attributes highlighted as harnessing individual differences in chess playing were memory, accurate analysis, quickness of perception, constructive imagination, and far-sighted combinations of chess pieces (Cleveland, 1907). Furthermore, factors such as memory, attention, high-level intellectual processes, imagination, will, and psychological type were the main psychological factors advanced as underlying chess talent (Djakow et al., 1927).

The first fully comprehensive work about chess psychology, however, was that carried out by Adriaan de Groot, a psychologist and a proficient chess player and international master who represented the Netherlands in the Chess Olympiad (de Groot, 1965). Thought processes elicited in chess were studied by presenting chess players with positions from actual games while asking them to choose the correct move and think aloud the process to reach the given choice. One important finding in the work by de Groot was that grandmaster players showed a striking superiority in selecting the strongest moves in a chess position, compared with lower-level players who tended to select poorer moves. Subsequent work has placed the focus on three main mental processes that are central in the domain of chess: perception, memory, and thinking.

## 3.1 Perception

These days computers hold an enormous advantage in chess playing over human opponents. It has now been more than twenty years since the Deep Blue machine defeated world chess champion Garry Kasparov in an epic battle in 1997 (Campbell, Hoane, & Hsu, 2002). Specific advantages of machines over humans in chess playing encompass a higher short-term memory capacity, speediness in individual calculations, the avoidance of errors, and analysis of the position free from emotional and instinctive constraints (Berliner, 1974; Shannon, 1950). The design of the first computers capable of playing chess entailed human-like attributes such as perception, problem evaluation, and decision-making with regard to the most likely moves leading to a solution of a given chess problem (Scurrah & Wagner, 1970; Simon & Barenfeld, 1969). Machines base their chess strength, however, on an efficient search within the chess tree of valid continuations arising from a given position (see Figure 3.1), which readily becomes massive after a few moves from both players (Shannon,

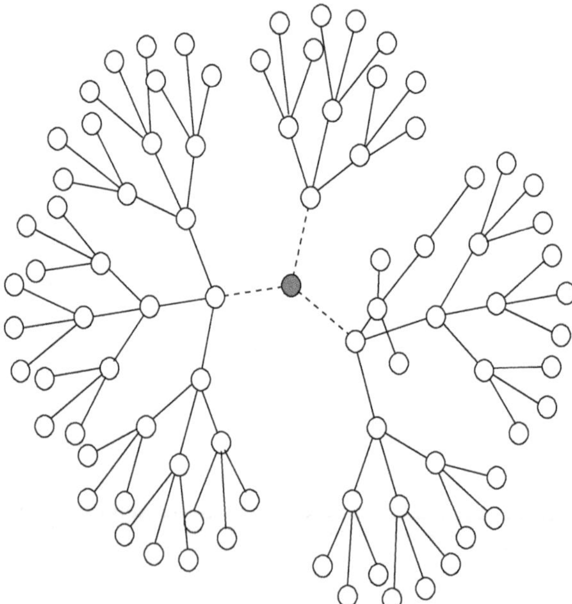

**Figure 3.1** The representation of a chess tree with 86 nodes begins in the black central node, which splits into three main variants (dotted lines); each successive node splits into three lower-level nodes, representing the alternative choices arising at each main variant

1950). In searching within the chess tree, an important aim of original chess software was in fact to narrow the search in a meaningful way (Berliner, 1974). Modern chess-playing computers in the last two decades, from Deep Blue to Alpha Zero, have excelled in this kind of behaviour by implementing relatively novel algorithms, such as quiescence search, iterative deepening, transposition tables, principal variation, and deep neural networks (Campbell et al., 2002; Silver et al., 2018). Hence, chess engines are highly resourceful at thinking, heuristic problem solving, and searching more speedily and more deeply within the chess tree, while being able to defeat the strongest human players in the world.

In contrast, humans appear to perceive the properties and particularities of a chess position without initially seeking the specific solution through the branches of the chess tree (Simon & Barenfeld, 1969). There is an interesting finding, which has been replicated in several studies (Vicente & de Groot, 1990). Human chess players are asked to reproduce a given chess position, with over twenty chess pieces, that has been seen beforehand for a relatively short time interval of between two and ten seconds. The accuracy in reproducing the given position is positively correlated with individual differences in the level of chess skill. When the task is repeated with the pieces placed randomly in meaningless chess positions, though, the correlation is still positive, albeit lower than with normal meaningful chess positions (Chase & Simon, 1973; Simon & Chase, 1973). It has been remarked, however, that the correlation might hardly be significant, because of the low sample and effect size, while stronger players may still perform better than weaker players (Gobet, 1998).

This crucial finding highlights the importance of basic perceptual processes in explaining the overwhelming advantage of stronger players over weaker players when solving a chess problem or judging a chess position. From this viewpoint, even though the structure of the search process might be very similar across chess players of varying strength, there will be differences in the possible paths suggested by the chunks stored in long-term memory (Simon & Chase, 1973). Indeed, when recalling a random distribution of pieces, chess players with higher levels of chess skill still perform better than chess players with lower levels of chess skill (Gobet & Waters, 2003; Goldin, 1979). Furthermore, recall performance in terms of random positions may also depend on the convergence of the pieces on the same squares. Chess masters showed better recall performance than class A and class C chess players when the chess pieces were located in the central squares rather than in the peripheral squares of the board (Reynolds, 1982). In addition, it has been reported that stronger chess players are also faster in recalling chess positions, even though the variability in response time across normal and random positions is unrelated to chess skill (Saariluoma, 1985).

Information-processing models have been used extensively to address the perceptual processes of chess information (Chase & Simon, 1973; Simon & Barenfeld, 1969; Simon & Chase, 1973; Simon & Gilmartin, 1973). Table 3.1 shows some of these information-processing models. For instance, the MATER model was designed to detect significant relationships between the pieces and squares in a given chess position when searching for a checkmate continuation (heuristic search). The PERCEIVER model was designed to gather information about the relationships among the pieces and squares in a chess position (information gathering). The combination of EPAM with MAPP accounts for the perceptual human processes used to remember and reproduce chess positions after examining them during a short time interval. The CHREST model, based on PERCEIVER and MAPP, considers four components and four main mechanisms: eye movements, information encoding and storing into long-term memory, learning from long-term memory, and information updating in the mind's eye component (Gobet & Simon, 2000; Waters & Gobet, 2008).

Table 3.1 *Information-processing models to explain perceptual processes in chess (EPAM = elementary perceiver and memorizer; MAPP = memory-aided pattern perceiver)*

| Model | References | Main goal | Postulates/components |
|---|---|---|---|
| MATER | (Simon & Chase, 1973) | Search for checkmate continuations. | Humans use information from a position and apply heuristic rules to select a reduced set of potential solutions for further consideration. |
| PERCEIVER | (Simon & Barenfeld, 1969; Simon & Chase, 1973) | Simulate human eye movements. | Gather information about meaningful relations between chess pieces related to chess rules, such as defending and attacking. |

Table 3.1 *Cont.*

| Model | References | Main goal | Postulates/components |
|---|---|---|---|
| EPAM | (Simon & Chase, 1973; Simon & Gilmartin, 1973) | Store tests of chessboard locations in a binary tree structure. The terminal nodes of the tree (leaves) store specific known patterns of chess pieces. | There is a single learning process and a single resulting chunk. The time needed to fulfil a given learning task is proportional to the amount of chunks used in completing the task. |
| MAPP | (Simon & Chase, 1973; Simon & Gilmartin, 1973) | Simulate the human processes to remember and reproduce chess positions that have been seen briefly. | This model comprises learning and performance components. The learning component uses the mechanism of EPAM; the performance component uses the mechanisms of PERCEIVER and EPAM. |
| CHREST | (Gobet & Simon, 2000) | Model learning and expertise. | This model comprises four modules: a simulated eye, a discrimination network that gives access to long-term memory, a short-term memory, and a mind's eye. |

The exploration of eye movements constitutes another central line of research addressing the visualization of chess positions and chess playing. Initial work on this topic investigated chess master's impressive efficiency in grasping the significance of a chess position. More skilled players might fix their eyes on the most salient pieces of the position, encode two or more chess pieces in a single eye fixation, cover larger chessboard areas – in sum, perceiving groups of pieces as single meaningful information units (de Groot, Gobet, & Jongman, 1996; Tikhomirov & Poznyanskaya, 1966). Further empirical studies have also indicated superior perceptual ability on the part of expert over non-expert chess players. Stronger expert players appear to make fewer eye fixations, albeit greater-amplitude saccades, than intermediate players. In addition, stronger players fix their eyes more frequently on empty chessboard squares and on the most relevant pieces that conform to a particular chess position. These findings suggest that, rather than encoding individual pieces, expert players have an advantage in encoding chess configurations or broad patterns incorporating several chess pieces (Charness et al., 2001). Expert chess players also tend to invest more time in studying the relevant squares for familiar solutions than the relevant squares for the optimal solution. These findings have been interpreted as supportive of the Einstellung effect, whereby one fixed idea prevents other ideas coming to mind (Bilalić, McLeod, & Gobet, 2008b).

The combination of the analysis of eye movements with the recording of brain activity has confirmed the remarkable advantage of experts over non-experts. This advantage relates to the more extensive knowledge base of diverse chess patterns. For example, studies applying the electroencephalography technique indicate that the chess players who are more proficient at attending to the relevant parts of the position present more brain activation in areas related to planning and decision-making, such as the prefrontal cortex. In contrast, chess players who are less proficient fix their eyes on larger spaces, which requires the processing of a larger amount of information. These players show more brain activation in areas related to the processing of visual stimuli, such as the occipital and parietal cortex (Silva-Junior et al., 2018). Moreover, findings from fMRI studies have indicated that chess experts are able to recognize a given meaningful chess position while ignoring trivial configurations quickly and efficiently, although this ability decreases with random patterns of meaningless chess positions (Bilalić, Kiesel, et al., 2011; Bilalić et al., 2010, 2012).

On the other hand, it has been suggested that the higher perceptual encoding ability of expert chess players might also depend on their higher chess experience and knowledge, rather than on a general perceptual or memory superiority (Kiesel et al., 2009; Reingold et al., 2001). From this viewpoint, the superior recall skills of expert over novice chess players for briefly presented chess positions are associated with the intrinsic disposition of the pieces on the board – i.e., the geometry, form, and colours of the chessboard and pieces, and the greater familiarity of experts with meaningful dispositions of chess pieces (Bilalić, McLeod, & Gobet, 2009; Schneider et al., 1993).

For example, an interesting study in this line of research adopts the diametrical terms of ‘underdeveloped’ or ‘developed’ chess patterns (Ferrari, Didierjean, & Marmèche, 2008). Underdeveloped chess patterns correspond to chess positions whereby the pieces are located in the chessboard rows 1, 2, 7, and 8, which may bear a lower strategic value. Developed patterns correspond to chess positions whereby the pieces are located in the chessboard rows 3, 4, 5, and 6, which may bear a greater strategic value. Ferrari and collaborators examined whether experienced players would be faster at encoding underdeveloped than developed patterns with a flicker paradigm and a recognition task in a two-experiment study. As expected, the findings indicate that more experienced players are faster at encoding developed patterns than at encoding underdeveloped patterns (three to four seconds), whereas novice players are slower (four to ten seconds) and invest an analogous amount of time in encoding both kinds of patterns. Expert players encode underdeveloped patterns – i.e., strategically unimportant ones – in a global and quick manner that allow them to focus instead on the strategically important developed patterns. Such a general strategy would be likely to allow for a more efficient application of analytical skills, which are, in turn, highly dependent on memory skills. Indeed, one of the fundamental elements during eye movements is visuospatial working memory (van der Stigchel & Hollingworth, 2018). The main target is selected before a saccade by the actual available content in visual working memory, supporting the view that perceptual and mnemonic processes intertwine closely in chess.

## 3.2 Memory

Memory is a central attribute in chess playing. Once the flavour of a given chess position or problem has been captured through perceptual processes, a representation in both short-term and long-term memory occurs (Gobet & Simon, 1996a, 1996c, 1998a; Goldin, 1978, 1979). Because of the inherent limitations of human memory when confronted with an increased amount of information, the memorization of chess positions and combinations can be extremely taxing on account of the high density of the chess tree (Figure 3.1). Hence, individual differences in mnemonic abilities are likely to relate strongly to individual differences in chess performance (Baddeley, Thomson, & Buchanan, 1975; Furley & Wood, 2016).

The stronger chess players hold a remarkable advantage over weaker chess players in recognition and recall processes tapping mnemonic structures (Goldin, 1978, 1979). There are four main models that account for the role of expert memory in chess and other domains (Gobet, 1998): the chunking theory (Chase & Simon, 1973); the search, evaluation, and knowledge theory (Holding, 1985); the long-term working memory theory (Ericsson & Kintsch, 1995), and the template theory (Gobet & Simon, 1996c). Table 3.2 summarizes these models by describing their main tenets and suggested degree of supportive evidence across five main dimensions: early perception; short-term recall and long-term encoding; modality of representation; long-term organization; and learning. About half the empirical available data were unsupportive of

Table 3.2 *Overview of theories addressing the role of expert memory in chess (LTM = long-term memory; SEEK = search, evaluation, knowledge) with the degree of supportive and unsupportive evidence (Gobet, 1998)*

| Theory | Reference | Main tenets | Supportive | Unsupportive |
|---|---|---|---|---|
| Chunks | (Chase & Simon, 1973) | Chess positions are encoded in memory as larger units than isolated pieces, including the relationships between pieces and empty squares – so-called chunks. Chess skill relies on the collection of chunks stored in LTM, allowing the recognition of very specific patterns. | 92% | 8% |
| SEEK | (Holding, 1985) | The three main elements in expert memory are search, evaluation, and knowledge. Search and evaluation draw on a broad knowledge base accumulated through years of experience. Chess positions and moves are selected from the sequences generated within a given chess tree by integrating search, evaluation, and knowledge. | 31% | 54% |
| LTM-WM | (Ericsson & Kintsch, 1995) | Chess positions are recovered from the 64-square chess-board defining a hierarchical structure that relates the chess pieces to each other and to their respective locations. The hierarchical structure is represented and working memory, allowing an accurate analysis and evaluation. | 38% (square) 62% (hierarchy) | 46% (square) 15% (hierarchy) |
| Template | (Gobet & Simon, 1996c) | The three main elements in expert memory are a database of chunks, a knowledge base, and a connection of the chunks to the knowledge base. Templates are larger chunks specifying the location of several pieces, and slots, eventually accommodating additional pieces and their locations. | 100% | – |

with high-level descriptions of chess positions with sentences such as 'Queen's gambit declined exchange variation-type position. White is conducting a minority attack. Black has defensive resources and some prospect of a kingside attack' (position 2). The recall of the chess positions improved when the description was presented before rather than after actually seeing them, which was deemed as supporting the view that high-level abstract knowledge is the main determinant of the ability to recall chess positions (Cooke et al., 1993). Further outcomes suggest that stronger players are better able to integrate familiar chess piece configurations into comprehensible schemes (Lane & Robertson, 1979), and that the quality of the best chess moves depends on evaluation rather than on recognition processes (Holding & Reynolds, 1982). Moreover, knowledge of the legal rules in chess exerts a more influential role than memory for chess positions at lower levels of expertise (Yoskowitz, 1991), suggesting that deeper levels of processing contribute to a better retention ability. This latter claim was additionally substantiated with forty-four novice chess players (Marmèche & Didierjean, 2001), suggesting that knowledge generalization leads to the construction of abstract solving schemas, which promote in turn a better retention of context-dependent elements. A comprehensive study with forty-seven children examined in addition whether chess skill related to the judgement of the more abstract aspects of a chess position (Horgan & Morgan, 1990). The findings point to the fact that a growing level of chess skill is associated with a more abstract representation of the features of the chess position. A more recent study also corroborates that, apart from encoding the basic elements of a given chess position, highly rated players tend to encode more abstract and semantic chess relationships (Gong, Ericsson, & Moxley, 2015). Taken together, these findings support memory as a by-product of cognitive-perceptual processes, which might rely on a rich network of verbal and evaluative information while increasing in importance towards more positional or complex chess positions (Pfau & Murphy, 1988).

Nevertheless, verbal rehearsal may have little influence on the memory for chess positions or the analytical processes involved in the selection of chess moves. This point of view highlights that a more suitable cognitive function to use in the explanation of individual differences in chess skill is likely to be embedded within long-term memory (Robbins et al., 1996). Similarly, an extension of the study by Holding and Reynolds (1982) reached the opposite conclusion regarding the role of pattern recognition (Schultetus & Charness, 1999). There was an obvious link between the quality of move selection and the subsequent recall performance, which necessarily implies a representation either in long-term memory or in a template structure. Furthermore, chess experts of a similar skill level albeit specialized in different chess openings were asked to solve problems within four kinds of chess positions arising from four different openings: Sicilian, French, Neutral, and Random (Bilalić, McLeod, et al., 2009). There was better performance in both remembering and solving problems concerning the respective area of specialization, indicating that knowledge of specific areas was more important than general problem-solving skills. This finding was argued to support the template theory hypothesis regarding the connection between templates with potential solutions and plans of action. Moreover, both the chunking and template theories contemplate the importance of abstract knowledge, whereby templates are deemed as conceptual prototypes that provide labels for characterizing extensive families of chess positions (Gobet, 1998; Gobet & Simon, 1996c).

When dealing with chess problem solving, however, it is important to bear in mind the role of judgement and decision-making, two crucial cognitive processes that are embedded in the broader behavioural dimension of thinking.

## 3.3 Thinking

Chess thinking has attracted considerable attention from expert chess players and scholars. For example, Nikolai Krogius suggests that there are two main ways of thinking when playing chess: concrete calculation and general assessment. Either thinking method might predominate depending on the specific moment and circumstances arising during a chess game (Krogius, 1976). A more recent chess training book suggests that chess playing comprises three main thinking strategies (Samarian, 2008): positional intuition; the ability to 'see' combinations; and the ability to calculate variations quickly and efficiently. Intuition differentiates master from average-level players because intuition predominates over logical thinking in chess (Kelly, 1985). In another view, emotional experience triggers the activation of thinking activity (Tikhomirov & Vinogradov, 1970). Moreover, other explanations of chess thinking refer to the term 'apperception', the application of unconscious principles for representing mental content and decision-making (Saariluoma, 1995, 2001).

Nevertheless, the first in-depth work about chess thinking was that carried out by de Groot between 1938 and 1943 with renowned chess players such as Alexander Alekhine, Max Euwe, Paul Keres, Reuben Fine, Salo Flohr, and Savielly Tartakower (de Groot, 1965). Apart from laying down the foundations for the psychology of chess on cognitive and experimental grounds, the work by de Groot acknowledges the role of individual differences. The central aim of this work was to characterize the thought dynamics in chess regarding its organization, methods, and operations, in line with Otto Selz's notion of thinking. More specifically, de Groot asked about the thought processes underlying chess skill, and about the singular ability of chess masters to spot the best moves usually overlooked by average chess players. These questions were examined through the verbalization of thought processes elicited when proposing the stronger chess move leading to the solution of real chess positions. There were twenty-two chess players in this study: the aforementioned six grandmasters, four masters, two female champions of the Netherlands, five strong experts, and five skilled players. All players adopted similar thinking and decision-making strategies regardless of their chess strength. Stronger players were able to recognize at a glance the essence and correct move of the proposed chess position, however. The thinking-aloud methodology furnished some understanding about chess thinking methods and chess styles, albeit encompassing a considerable degree of inter-individual variability and 'individual peculiarities and idiosyncrasies' (de Groot, 1965: 313).

Human judgement and decision-making in a variety of intellectual domains have been approached as pivoting on two thinking processes. Both processes have been referred to in many different ways. For example, the seminal work by Keith Stanovich and Richard West labels the processes as system 1 and system 2 (Kahneman, 2011; Stanovich & West, 2000). System 1 is a quick, automated process, operating effortlessly; system 2 is a slow, demanding process, usually involving complex computations and mental concentration. When judging a situation implying decision-making, both systems operate, at times in a close partnership. Whereas system 1 generates patterns of general ideas, system 2 builds specific methodical thoughts. System 1 and system 2 have also been termed 'intuitive' and 'analytical' components, operating together in human judgement and decision-making (Betsch & Glöckner, 2010). This systematic approach to thinking should also operate in the intellectually demanding endeavour of chess playing.

Some thinking skills of expert chess players are certainly astonishing. Skilled child players are better able than non-chess-playing adults to predict their own performance even in a domain outside their own field of expertise (Horgan, 1992). Stronger chess players are also more accurate than weaker players at estimating a player's strength from observing a particular self-created chess position (Reynolds, 1992). In addition, the estimation error of this prediction decreases when they see the subsequent moves generated in the position. The most striking advantage that expert players have over novices bears upon finding the best move in a given chess position, however (de Groot, 1965). This superiority is most likely to lie well beyond a simple advantage in perceptual or memory abilities, such as several high-level cognitive processing abilities that account more precisely for individual differences in chess skill. Some of these processes are forward-visual evaluation (Holding & Pfau, 1985), information processing (Horgan, Millis, & Neimeyer, 1989), pattern recognition (Schultetus & Charness, 1999), planning performance (Unterrainer et al., 2006), language (Nippold, 2009; Pfau & Murphy, 1988; Vasyukova, 2012), geometric and numerical skills (Ferreira & Palhares, 2008), and visuospatial and verbal encoding (Bachmann & Oit, 1992; Wagner & Scurrah, 1971). For example, chess players at higher levels of chess skill might be more skilful in foreseeing future positions by discerning non-redundant similarities, alternative goals, and more candidate moves in a given chess position (Holding & Pfau, 1985; Horgan et al., 1989).

Why are there such remarkable individual differences in thinking efficiently about chess information? The scientific literature has put forward two main views anchored in the aforementioned system 1 (intuition) and system 2 (analytical) processes, which at the same time has been the object of a vigorous debate. The processes have been termed 'pattern recognition' and 'search', respectively. In one view, pattern recognition wields a dominant role and search a secondary role in chess thinking (Gobet & Simon, 1996b). This claim is largely based on the outcomes from simultaneous chess games between 1985 and 1992 played by the chess world champion at that time, Garry Kasparov, against national teams consisting of four to eight players. Under these conditions, Kasparov endured serious time constraints that prevented him from being involved in slow search. On the other hand, he was allowed to analyse, with the aid of a computer, 100 games from each of his opponents prior to each tournament. Because Kasparov played at a very high level under these circumstances, this was interpreted as supportive evidence for the prevalence of pattern recognition compared with search.

Conversely, and in the view of the search approach, pattern recognition has been judged as inadequate to account for individual differences in chess skill, casting serious scepticism over the guiding premises of the pattern recognition theory (Holding, 1992; Holding & Pfau, 1985). From the search viewpoint, individual differences in chess thinking and chess skill are likely to depend on slow search processes, as portrayed by the SEEK (search, evaluation, and knowledge) theory. A crucial postulate of this theory is that the selection of chess moves proceeds by anticipating consequences with chess trees that are smaller and more discriminating, however, than chess trees built by machines. Moreover, the SEEK theory claims that there are notable individual differences in chess skill regarding search, evaluation, and knowledge, which remain with time constraints in fast chess and are not accounted for well solely by the pattern recognition approach.

Thinking in chess is largely influenced by the available thinking time. Time is also a central key distinction between the system 1 and system 2 thinking processes. Therefore, comparing chess performance in fast and slow chess playing is an appealing approach to study chess thinking. There are two main chess-playing modalities regarding thinking time: fast (blitz) and slow games. Typically, chess players endure severe time constraints in fast blitz games, whereas they have a higher amount of available thinking time in slow games. Time constraints, therefore, might foster fast pattern recognition processes, albeit increasing in turn the probability of committing devastating errors (Sigman et al., 2010). Moreover, time constraints might influence strategic behaviour and thinking during chess playing. A recent experiment with a large dataset of internet blitz chess games found that players changed their playing strategy when confronted with stronger opponents (Fernandez-Slezak & Sigman, 2012). Chess players adopted a prevention mode when playing stronger opponents, which involved investing a higher amount of thinking time together with greater move accuracy. This strategy did not imply increasing the probability of winning, however, because this extra time did not lead to significant improvements in the position. Increasing the probability of winning against stronger opponents would require a playing style more in accordance with that adopted when playing against an opponent with equivalent chess strength.

Several stimulating studies have addressed the dual view of chess thinking by studying time constraints during chess playing, which have provided either supportive or unsupportive evidence for both pattern recognition and search as the predominant thinking processes in chess. An early study with six US players compared the quality of chess moves during standard games lasting forty moves in ninety minutes with blitz games lasting only five minutes for each player (Calderwood, Klein, & Crandall, 1988). Blitz time constraints hampered the quality of chess moves to a greater extent for weaker than for stronger players, which was considered as supportive of the pattern recognition approach. Furthermore, four chess players of different levels of expertise – grandmaster, international master, candidate master, and class B player – were asked to solve complex chess positions that required search thinking, and also to solve rapid decision-making, memory, and practice tasks (Campitelli & Gobet, 2004). The findings supported that search performance increases linearly with chess skill, in accordance with the SEEK theory. These findings also supported the pattern recognition theory in the memory and fast decision tasks, however. This study highlights that search might be undertaken occasionally in chess, whereas selective search and pattern recognition are also important for both fast and slow decision-making.

There is indeed evidence that bolsters the pattern recognition view. For example, thirty-four US chess players were challenged to predict the moves of a chess game played by experts (Klein & Peio, 1989). Stronger players were more accurate than weaker players at predicting the moves of the game while also being more likely to advance the correct option as the first one to be considered. This set of findings was believed to reflect an underlying pattern recognition model as the main explanatory process of individual differences in chess skill. The effect of finding the best choice as the first one considered was replicated in another study that presented sixteen chess players with one middlegame position and three endgame positions (Klein et al., 1995). That experts are likely to rely on the generation of feasible options as the first ones considered was consistently corroborated. Moreover, similar verbal protocols to those used by de Groot were applied in the analyses of two chess positions with a sample of twenty-two intermediate and master chess players recruited from chess tournaments in Australia (Connors, Burns, & Campitelli, 2011). Masters had faster search abilities than intermediate players, through being better at generating more board positions resulting from the consideration of potential candidate moves. Masters did not search more deeply than intermediate players did, however. Thus, this study underlines the importance of fast pattern recognition for chess skill despite the fact that chess masters displayed faster and deeper searching abilities than intermediate players.

Apart from studying chess players making decisions about chess positions, support for the pattern recognition approach has emerged from other kinds of data. For instance, a positron emission tomography (PET) study with blindfold chess compared the activation of brain areas in three kinds of tasks: attention, memory, and problem solving. These findings suggest that visuospatial representations of chess piece configurations are in fact prelearned chunks stored in long-term memory while implying highly automatic processing, which appears to support to a great extent the pattern recognition approach (Saariluoma et al., 2004). Another study with archival databases involving thirteen blitz chess tournament outcomes from the Netherlands, the United States, and Australia also confirmed the pattern recognition view (Burns, 2004). Fast processes such as recognition accounted for most of the variance in overall chess skill, whereas chess skill differences equalized when blitz games were being played, albeit to a lesser extent at higher levels of chess skill. In addition, although searching might be important for chess performance, it might also remain constant beyond a certain level of high skill.

Moreover, there is alternative evidence about the pattern recognition approach, emphasizing instead the importance of search processes in explaining individual differences in chess skill. For example, the findings from the simultaneous chess games played by Kasparov, which support the pattern recognition predominance in chess (Gobet & Simon, 1996b), have been challenged by those advocating instead for the search view. The main argument hinges on the fact that the outcomes of games played between humans and machines indicate a stronger advantage for machines when the available thinking time is decreased. Such a trend emphasizes that human ability in search decreases with more stringent time constraints. Conversely, increasing the available thinking time favours search to a greater extent while enhancing chess-playing strength

(Lassiter, 2000). Moreover, the number and magnitude of chess blunders committed by twenty-three chess grandmasters were analysed by contrasting three kinds of chess games: slow, rapid, and rapid blindfold (Chabris & Hearst, 2003). A higher number of errors and larger blunders arose more often in rapid chess than in slow chess, whereas blunders were very similar in both respects in rapid and rapid blindfold chess alike. These findings were taken as opposed to the fast pattern recognition view, because grandmasters made a higher number of errors and irrecoverable mistakes with thinking time constraints. Similarly, a study comparing the games in slow and fast games of grandmasters, international masters, FIDE masters, and untitled players, with seventy-five players in each group, yielded analogous findings (van Harreveld et al., 2007). Time constraints resulted in a lessened predictive ability of individual differences in chess skill in blitz games than in slower games, suggesting that slow search thinking processes are important for both the stronger and the weaker players.

A more recent study involving blitz games examined the time invested in chess moves apart from chess blunders while contrasting individual differences in chess skill (Chang & Lane, 2016). The two main findings contradict the fast pattern recognition approach. First, stronger players were more prone than weaker players to spend a considerable amount of time on a few moves. Second, the stronger players committed fewer blunders than weaker players. These findings support the notion that deep calculation is indeed still possible in fast chess, and that individual differences in the ability to search contribute meaningfully to individual differences in chess skill. The pattern recognition approach together with the concept of chunks has also been analysed in depth on theoretical grounds (Linhares & Freitas, 2010). With applied examples, this conceptual study reflects on the seminal work about perception in chess (Chase & Simon, 1973), by proposing the alternative ideas of experience recognition and analogies as the crucial elements of thinking in chess. In the view of Alexandre Linhares and Anna Freitas, chunks that form the simplest building blocks for more complex patterns should necessarily convey semantic information. Their argument appears to be in direct opposition to the pattern recognition approach. It questions the link between knowledge of very specific patterns (stored in long-term memory) and the actual selection of moves, and aligns better with findings that emphasize the importance of search processes during chess playing (Chabris & Hearst, 2003; McGregor & Howes, 2002).

Perhaps a more reasonable perspective is one that highlights fast pattern recognition and slow search as both being equally important for chess skill. Harmonizing the two main viewpoints about the underlying thinking process of chess skill would imply that, with more thinking time, there might be greater chances of recognizing more patterns and applying search thinking to them (Chabris & Hearst, 2003). A very similar conclusion was also reached in another classical chess experiment with seventy-one chess players, who had to find the best moves to fifteen chess positions while thinking aloud (Moxley et al., 2012). Experts and less skilled players alike benefited from extra deliberation independently of the difficulty of the respective chess problem. Thus, both intuition (pattern recognition) and deliberative thinking (search) were deemed as mutually beneficial for chess skill. In another view, pattern recognition (i.e., chunking theory) in fact embraces search processes as well.

Fernand Gobet, the leading researcher in the field of chess psychology, admits that it is a recurrent misunderstanding of chunking theory to presume it to be the unique or principal underlying factor to explaining individual differences in chess skill. A more integrated view holds that both the recognition of chunks and the exploration of candidate moves and their consequences underlie individual differences in chess playing skill (Gobet, 1997). Gobet argues that several of the premises raised by the SEEK theory (Holding, 1992) fail to provide evidence against the chunking theory. More specifically, the chunking theory is argued to provide the mechanisms of pattern recognition as an explanation of the search and evaluation of chess positions. Chess skill would therefore depend on large amounts of regular patterns of chess piece configurations stored in memory, together with selective search strategies within the chess tree. Knowledge structures gained through the acquisition of expertise in long-term memory would also increase short-term memory capacity when confronted with specific chess situations (Gobet, 1997; Gobet & Simon, 1998b).

Chess thinking entails an intensive and complex combination of perceptual, memory, and thinking processes. In addition, chess thinking is influenced by a variety of psychological factors underlying individual differences. Moreover, chess playing even at an amateur level implies learning a considerable amount of domain knowledge, eventually extending over a lifetime to achieve a master level. Individual differences in brain functioning, intelligence, and personality are strongly associated with several learning processes in life domains such as education, work, and health. It is conceivable that they should play a role as well in learning and playing chess. Compared with other domains, however, the available scientific evidence linking individual differences in brain, intelligence, and personality with chess playing and chess skill is scanty, albeit encouraging.

Individual differences highlight the variability in behaviour occurring between and within individuals, and between groups of individuals sharing a particular characteristic, such as when comparing novice with expert chess players (Ackerman, 2014; Andrés-Pueyo, 1997; Colom, 2005). The next chapter delves into the individual differences approach to the psychology of chess, by analysing three main issues: the characterization and appraisal of individual differences; how individual differences can be addressed with a comprehensive model of adult intellectual development; and the controversy over heredity versus environment.

# 4

# Individual Differences

Chess has served as a kind of *Drosophila* model for cognitive psychology.

The previous chapter outlined three central cognitive processes involved in chess playing: perception, memory, and thinking. Information-processing models (Simon & Barenfeld, 1969; Simon & Chase, 1973; Simon & Gilmartin, 1973), the template theory of expert memory (Gobet, 1998; Gobet & Simon, 1996c), and system 1 and system 2 thinking processes (Kahneman, 2011; Stanovich & West, 2000) have provided a wealth of theoretical foundations and empirical findings useful for understanding how humans behave in an intellectually demanding environment such as chess.

This body of knowledge emphasizes that there is a considerable degree of variation in perceptual, memory, and thinking processes, however, eventually leading to meaningful individual differences in chess skill. As highlighted in Chapter 3, the studies about the perception, memory, and thinking of chess players point out that stronger players perform better than weaker players at several attributes useful for chess. These attributes consist of encoding chess positions into larger perceptual chunks (Chase & Simon, 1973; Gobet & Simon, 1996a, 1996c; Horgan, 1992; Lane & Robertson, 1979), memorizing chess positions, and discovering semantic networks among pieces and chunks (Frey & Adesman, 1976), recognition accuracy (Goldin, 1979), and evaluative judgement abilities (Holding & Pfau, 1985; Horgan et al., 1989; Klein & Peio, 1989).

Chess has also been suggested as an ideal model environment for the study of individual differences in the acquisition and development of expertise (Chabris, 2017; Charness, 1992; Gobet, 2016; van der Maas & Wagenmakers, 2005). In this view, studying individual differences in chess can contribute to unearthing how, when, and why humans differ when involved in a complex intellectual endeavour. Differential psychology is a useful conceptual and methodological approach to address such a different class of questions from those addressed so far from a cognitive psychology approach (Revelle, Wilt, & Condon, 2011). A considerable body of research within several fields and subfields of neurosciences and behavioural sciences advocates for the crucial significance of individual differences in explaining human behaviour. Several studies point out that there are consistent associations between brain anatomy and functioning and inter-individual differences in motor behaviour and learning, perception, higher-level cognition, and intelligence and personality (Kanai & Rees, 2011). Furthermore, individual differences in core psychological constructs such as intelligence, personality, and vocational interests account for meaningful variations in a vast array of behaviours that are readily observable and of cardinal importance for contemporary society (Lubinski, 2000).

This chapter is divided into three main sections. The first, on the characterization and appraisal of individual differences, describes the object of study and kinds of questions addressed by differential psychology, the definition and classification of psychological traits, and the main research designs in the field. The second section, on the individual differences in chess, introduces the intelligence as process, personality, interests, and intelligence as knowledge theory (PPIK) as an appropriate framework for integrating theoretical assumptions and contrasting hypotheses regarding the interrelationships of individual differences in psychological traits and chess skill. Moreover, it describes the Amsterdam Chess Test (ACT), a comprehensive psychometric device useful to evaluate individual differences in chess skill. Finally, the third section, on heredity versus environment, describes the most contentious issue within differential psychology – and probably within the social sciences. This section summarizes the main tenets of behavioural genetics, and some of the most remarkable findings in the two broad areas of intelligence and personality. Moreover, it highlights the relevance of the topic for chess research, particularly when addressing the talent versus practice controversy in accounting for the development of chess skill.

## 4.1 Characterization and Appraisal of Individual Differences

Individual differences were probably first studied thoroughly back in the sixteenth century with the groundwork laid by the Spanish philosopher Juan Huarte de San Juan (Huarte de San Juan, 1593). Huarte de San Juan was one of the first authors to build a psychological framework based in biology. Two of his main postulates centred on the adaptation of education to the kind of skill possessed by each individual (Velarde-Lombraña, 1993) and on the role of individual differences in intelligence (Gondra, 1994). Nearly 300 years later, one of the most influential works for the development of differential psychology was that carried out by Francis Galton in the nineteenth century (Galton, 1869). Galton applied Charles Darwin's evolutionary principles to human behaviour in an attempt to explain the foundations of intellectual abilities through the measurement of very basic sensory processes.

Contemporary differential psychology enquires into the variability and regularities in behaviour across individuals and groups. It also focuses on differences in psychological attributes and dimensions, and on individual variability regarding changes in behaviour across different moments and situations (Anastasi, 1937). From a comprehensive theoretical framework such as evolutionary psychology, individual differences are important for three main reasons. First, individual differences are very well documented. Second, individual differences have an important heritable component while exhibiting considerable stability over time. Third, stable individual differences have important consequences for evolution and adaptation to the circumstances encountered in real life (Buss, 2005, 2009).

Individual differences can be classified into inter-individual, inter-group, and intra-individual differences (Ackerman, 2014; Andrés-Pueyo, 1997; Colom, 2005). Inter-individual differences characterize differences between individuals (e.g., differences in verbal aptitude in a group of youngsters). Inter-group differences characterize differences between different groups (e.g., differences in academic achievement between males and females). Intra-individual differences characterize differences within individuals elicited during maturation or ageing, or when going through some kind of external intervention (e.g., the change in foreign-language proficiency throughout secondary education). Intra-individual differences also characterize differences between distinct attributes within individuals that can be evaluated with standard psychometric instruments (e.g., differences in performance between verbal and numerical abilities). Another kind of individual differences can be termed inter-individual differences in intra-individual change, which characterize the differential performance of individuals when exposed to some kind of intervention or when enduring some kind of change throughout their lives (Ackerman, 2014).

In general, individual behaviour manifests itself in a consistent and stable manner. Individuals are consistent and stable because they tend to display behavioural regularities across different situations and moments in time. Both consistency and stability are intimately related, with four main kinds of trait consistency being typically identified in the literature. Intra-individual differences in consistency (i.e., within trait dimensions), and ipsative consistency (i.e., between trait dimensions), centre on individual changes over time. Mean-level consistency centres on whether groups change in trait dimensions over time. Rank-order consistency centres on the relative distribution of individuals within groups. For example, it has been suggested that the rank-order consistency of personality traits increases progressively towards the age of sixty (Caspi, Roberts, & Shiner, 2005; Roberts & DelVecchio, 2000). Moreover, an earlier study suggested that the intra-individual consistency of traits might be hierarchically organized, being higher for intelligence, moderate for personality, and lower for narrower traits such as self-esteem and life satisfaction (Conley, 1984). With regard to intelligence, a longitudinal analysis of psychometric intelligence carried out at ages eleven and seventy-seven supported robust intra-individual consistency in intelligence (Deary et al., 2000). With regard to personality, a common phenomenon such as mate selection (i.e., assortative marriage) is also related to intra-individual consistency in middle adulthood (Caspi & Herbener, 1990). Furthermore, mean-level consistency is associated with the similarity between two situations and the personality of the individual (Sherman, Nave, & Funder, 2010).

Psychological traits are the narrower building blocks that constitute the basic architecture of differential psychology. Psychological traits describe the psychological attributes of interest concerning inter-individual variability in behaviour. There are at least ten useful qualities that are applicable to the identification of psychological traits (Eysenck & Eysenck, 1985; Zuckerman, 1991).

1. Psychological traits are useful dispositions for describing people and predicting human behaviour.
2. Psychological traits are concepts or constructs with a scientific foundation useful for describing individual differences.
3. Psychological traits are useful to place individuals within a continuous psychological dimension.
4. The responses to psychological testing characterize people's traits, and they can also help to elucidate a given psychological theory.
5. Innate and environmental factors both determine psychological traits, though all traits have a moderate level of heritability.
6. Psychological traits have meaningful biological correlates, and consistent associations with physiological subsystems embedded in the nervous system.
7. The interaction between psychological traits and specific situations produces mental states.
8. Psychological traits show consistent intra-individual stability.
9. Psychological traits are universal and identifiable with factor analysis by sampling behaviour across different methods, sexes, ages, and cultures.
10. Psychological traits derived from humans parallel behavioural traits in non-human species.

There are literally thousands of psychological traits amenable to a formal description with human language. Over 17,000 words in the English language were selected to describe human personality in one of the earliest studies about this topic (Allport & Odbert, 1936). For instance, multidimensional scaling statistical techniques applied to English-language terms yielded a structure as shown in Figure 4.1. The traits that are more beneficial for intellectual activities run from the most desirable at the top right corner, to the less desirable at the bottom left corner (Rosenberg et al., 1968). Figure 4.2 shows a potential classification of psychological traits into the two main broad domains of intelligence and personality (Andrés-Pueyo, 1997). Performance, ability, and skill are broad dimensions classified into the intelligence domain. Constitution, character, temperament, and personality

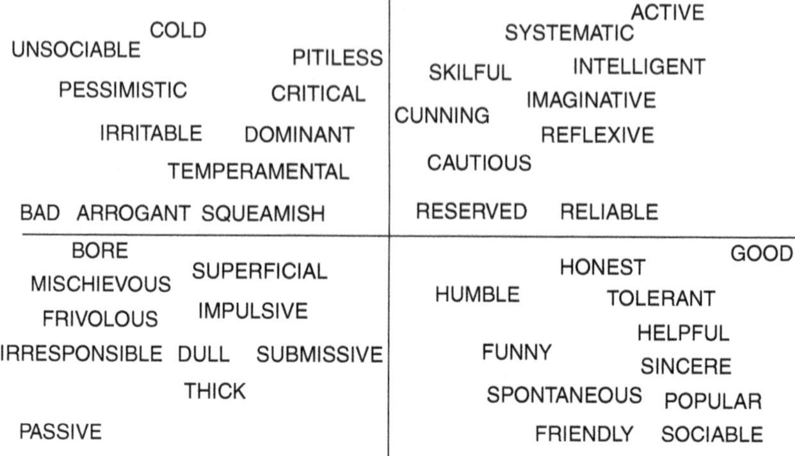


**Figure 4.1** A structure of personality impressions from a multidimensional scaling approach (Rosenberg, Nelson, & Vivekananthan, 1968)

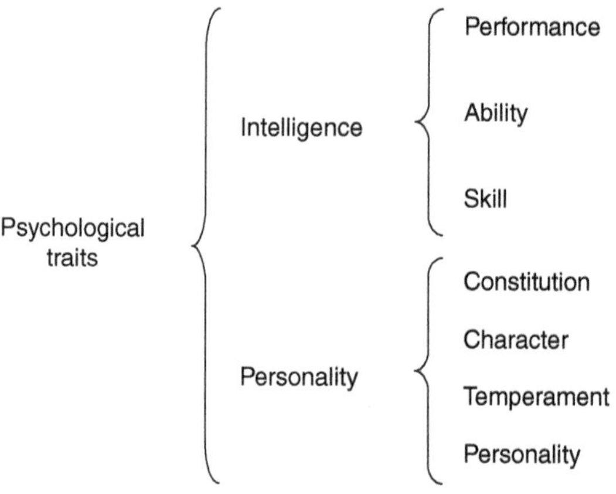


**Figure 4.2** A simple classification of psychological traits into two broad dimensions: intelligence and personality

are broad dimensions classified into the personality domain. Performance is a quantitative outcome accorded a criterion, which meets the responses given by an individual to a cognitive task (e.g., academic achievement, a college admission test such as the SAT). Ability is the level of potential performance reached in an assortment of specific behaviours (e.g., the intelligence quotient, the Elo chess rating). A skill is a set of competences exhibited when completing a task. In essence, a skill is the development of ability through practice or training (e.g., numerical, sport). Constitution is the biological structure of an individual in static terms, such as anatomical tissue, and dynamic terms, such as hormonal and biochemical functioning (e.g., height, second to fourth digit ratio). Temperament encapsulates the collection of emotional characteristics of behaviour somehow expressing the genetic determinants of personality (e.g., neuroticism, impulsivity). Character is the combination of traits related to attitudes, values, and feelings guided by culture (e.g., attitudes, beliefs). Finally, personality is the combination of factors related to constitution, temperament, character, and (for some authors) intelligence.

The measurement and evaluation of individual differences can be undertaken at three distinct levels of analyses underpinning distinct theories and measurement systems (Colom, 2005), as shown in Figure 4.3: the trait level, the

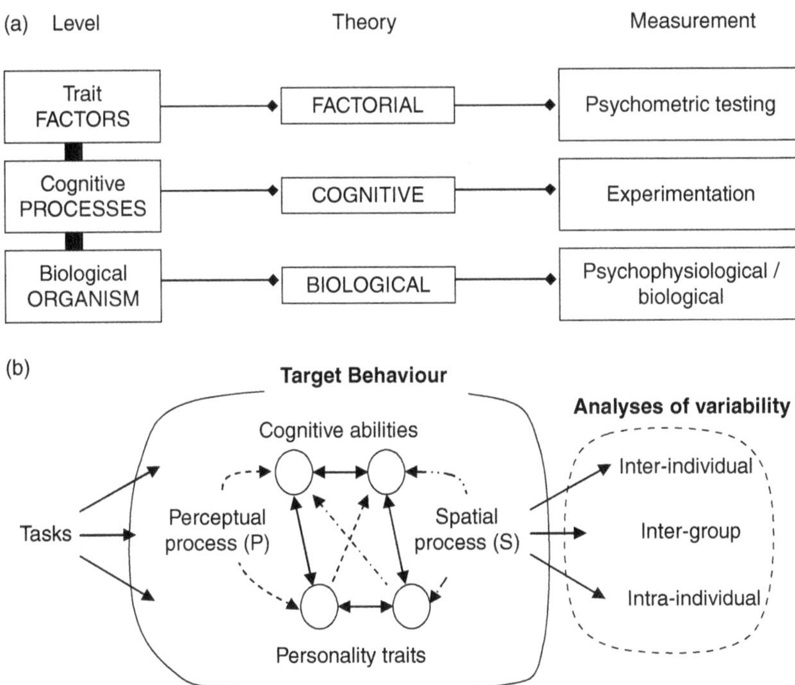


**Figure 4.3** There are different levels of analysis and measurement in differential psychology; the levels of analyses (traits, processes, and biological) can be combined to analyse the variability in a given target behaviour

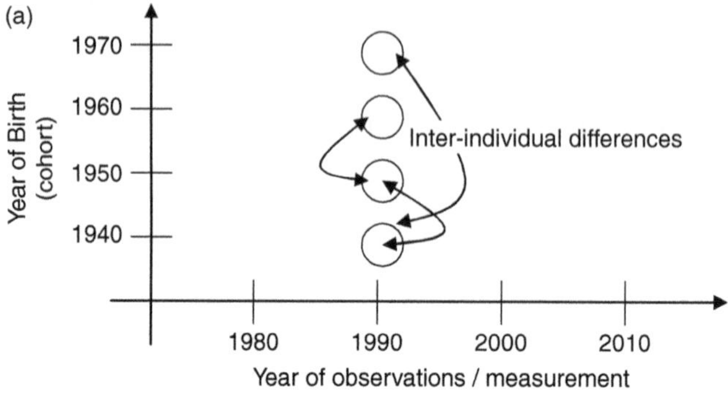


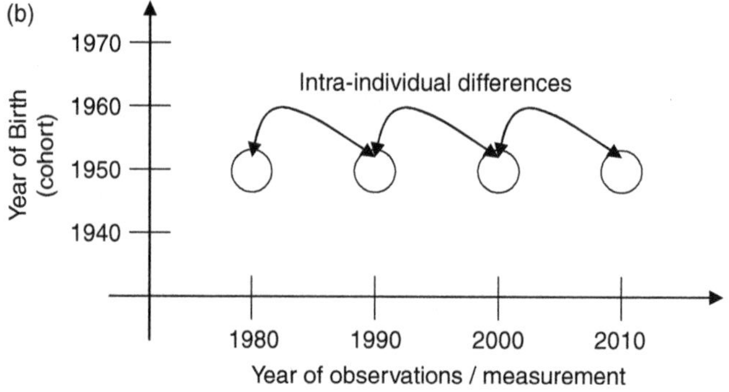


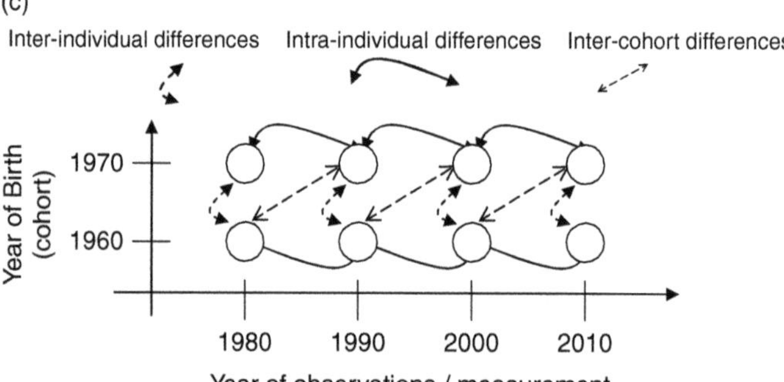


**Figure 4.4** Cross-sectional, longitudinal, and sequential research designs to evaluate inter-individual variability

cognitive level, and the biological level. The foundations of each level of analysis involve factorial, cognitive, or biological theories and their corresponding hypotheses. The associated measurement systems correspond to psychological testing, experimentation, and psychophysiological or biological measures. The variability in behaviour (i.e., inter-individual, inter-group, and intra-individual differences) can therefore be evaluated by combining different personality and intelligence traits at any level of analysis.

From a developmental approach, individual differences can be examined with three differentiated research designs: cross-sectional, longitudinal, and sequential (Figure 4.4). In a cross-sectional design, individuals of different ages or the same age are observed at the same temporal point. This design is particularly useful for providing information about inter-individual differences. With a longitudinal design, it is possible to observe the same individuals during a certain period and provide information about intra-individual differences. Finally, a sequential design combines the characteristics of both cross-sectional and longitudinal designs. In a sequential design, participants from different ages are selected and evaluated. This latter design allows the assessment of inter-individual differences as potential predictors of the intra-individual change, or whether intra-individual changes vary over time (Little, Schnabel, & Baumert, 1998).

## 4.2 Individual Differences in Chess

Chess is a demanding intellectual activity with systematic and universal foundations. It involves an intensive application of broad psychological attributes, such as perception, memory, and thinking. In addition, it subsumes an extensive amount of theoretical and practical knowledge about openings, endgames, strategical and positional characteristics, tactical combinations, and specific analytical and thinking methods. Learning and mastering this large body of knowledge requires a relatively long time, and also entails a great motivational effort. The inter-individual variability in these factors and processes is noteworthy, which determines to a great extent individual differences in chess performance and chess skill. Together with chess thinking methods, the variability in chess performance and chess skill have been considered to pivot on the connections between emotional, motivational, and cognitive processes (Tikhomirov & Vinogradov, 1970).

The intelligence as process, personality, interests, and intelligence as knowledge theory (PPIK) provides a comprehensive framework by which to integrate these individual differences, in accordance with both theoretical assumptions and empirical findings (Ackerman, 1996; Ackerman & Heggestad, 1997). Several empirical studies support the interplay of individual differences in ability, personality, and interests in the structure and development of intellectual performance (Hambrick, Meinz, & Oswald, 2007). The PPIK theory partly stemmed from three crucial inconsistencies concerning the assessment of adult intelligence:

1. the manifest contrast in the meaning of intelligence assessment for children and adults with intelligence quotient (IQ) tests;
2. the discrepancies in the poor performance of the elderly in measures of cognitive abilities when compared with their daily functioning; and
3. the correlations below 0.60 between measures of cognitive abilities and educational and job performance.

Aiming to override these limitations while providing a more straightforward theory of adult intellectual development, Phillip Ackerman has outlined a comprehensive and sound basis for evaluating the nature and interrelationships of individual differences during adult development. The PPIK theory is built on extant models of intelligence (Carroll, 1993; Cattell, 1987; Vernon, 1950), personality (Eysenck, 1952; McCrae & Costa, 1997), and vocational interests (Holland, 1959, 1996). Moreover, the PPIK framework provides a robust empirical background based on meta-analytic findings (Ackerman & Heggestad, 1997). The four main broad components of the PPIK theory are intelligence as process (fluid intelligence, *Gf*), personality, interests, and intelligence as knowledge (crystallized intelligence, *Gc*). Each of these broad components comprises narrower traits that can be evaluated when framed within a specific intellectual domain such as chess. Figure 4.5 shows a tentative PPIK representation adapted to the chess domain.

The four broad components of the PPIK theory are depicted in bold face, and hierarchically linked to their respective narrower traits with one-headed arrows. The narrower traits associated with intelligence as process (*Gf*) and interests are as suggested by the PPIK theory but for the long-term memory factor, and for the visual abilities factor, which replaces the spatial rotation factor suggested by Ackerman. These factors could be encapsulated under the umbrella of information-processing factors, which tend to decline with ageing. Discontinuous double-headed arrows represent correlations between the factors across the four main PPIK components. Realistic and investigative interests are associated with intelligence as process (*Gf*), while investigative and artistic interests are associated with intelligence as knowledge (*Gc*). The personality and intelligence as knowledge (*Gc*) components incorporate substantial changes from those indicated by the PPIK theory in line with the available findings in the chess domain. The personality component includes motivation, will power, and emotional regulation, because individual differences in these traits relate to chess skill in some studies. Openness and typical intellectual engagement were already suggested in the initial formulation of the PPIK theory. The intelligence as knowledge (*Gc*) component includes some of the most relevant elements that conform to the knowledge base of chess players.

For example, there are a huge range of chess writings dealing with the great variety of chess openings, their main lines, and their variants and subvariants.

This stage of the game is particularly important, because playing it well has a remarkable influence on the course of the game. An estimation of the opening knowledge of chess players across different levels of expertise suggests that chess players at the master level are able to memorize about 100,000 opening moves (Chassy & Gobet, 2011). Other core concepts contributing to the knowledge weaponry of chess players are positional and tactical principles, which are readily taught to beginners in chess-playing initiation courses. On the other hand, endgames correspond to the final stage of a chess game, when there are very few pieces remaining on the board. Endgames raise many specific schemes and subtleties, demanding thoughtful and intensive study in order to master them. Furthermore, Figure 4.5 indicates that the PPIK theory can be used to address both the structure and development of chess expertise (Ericsson et al., 2006; Gobet, 2016). In this view, the structure of chess expertise can be examined by looking at the pattern of traits most determinant of chess skill (e.g., inter-individual differences, inter-group differences), whereas the development of chess expertise can be examined by looking at the characteristics and properties of the change over time in those traits (e.g., intra-individual differences, inter-individual differences in the intra-individual change).

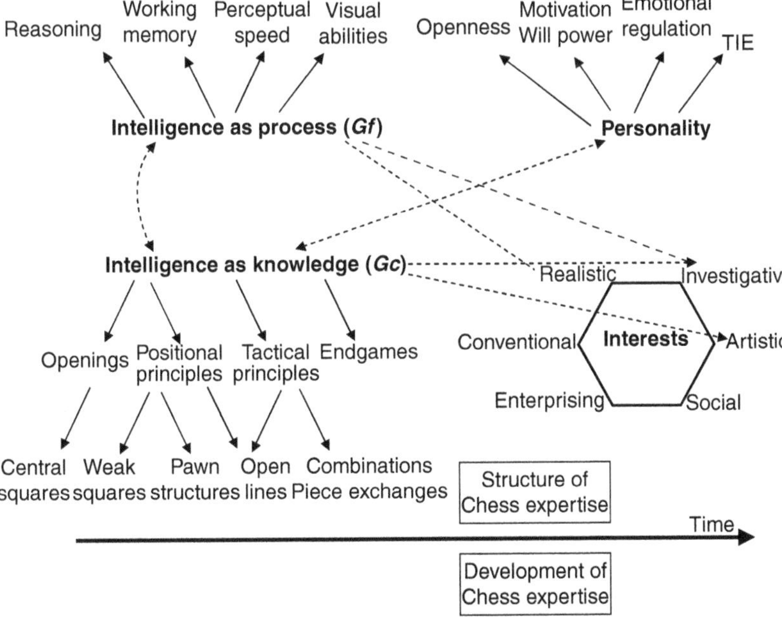


**Figure 4.5** The PPIK theory applied to the chess domain (intelligence as process, personality, interests, intelligence as knowledge: Ackerman, 1996); *Gf* = fluid intelligence; *Gc* = crystallized intelligence; TIE = typical intellectual engagement

Some of the links across the broad factors suggested by the PPIK framework have been advanced for instance in chess training books. For example, a recurrent question is whether these factors influence the choices that chess players make between valid continuations that are seemingly equivalent at first glance (Kotov, 1971; Samarian, 2008). Furthermore, earlier scientific works about the psychology of chess have also highlighted the importance of the factors embedded in the PPIK theory. Several general features of chess playing might influence individuals in a similar way in the emotional plane (Tikhomirov & Vinogradov, 1970). On the other hand, personal and temperamental differences on the part of chess players have been suggested as meaningfully shaping chess-playing style (Cleveland, 1907). Moreover, high-level intellectual processes such as combination power, finding logical regularities, and reaction time in relatively simple tasks, such as checking calculations, have been advanced as causal factors of chess talent (Djakow et al., 1927). In the work about thought processes in chess (de Groot, 1965), chess positions were presented to chess players with varying skill levels (e.g., grandmasters, masters, experts, and unskilled players). De Groot suggests that subtle systematic individual differences concerning problem-solving strategies arose, particularly when players were confronted with difficult chess positions. More specifically, de Groot suggests several areas, broad traits, and psychometric tests whereby individual differences should arise regarding the structure and development of chess talent, which can readily be identified within the PPIK framework:

1. higher scores in tests of spatial abilities;
2. higher scores in tests of verbal abilities;
3. an extensive knowledge base;
4. an outstanding learning ability from past experience;
5. the ability to test novel hypotheses;
6. deep motivation for trying and analysing;
7. an innate disposition to integrate thinking, playing, and fighting; and
8. unknown personality factors;

Because chess implies an intensive application of high-level cognitive processes (de Groot, 1965; Djakow et al., 1927), chess research has typically placed the greatest emphasis on the PPIK components of intelligence as process (*Gf*) and intelligence as knowledge (*Gc*). Intellectual abilities have been repeatedly put forward when seeking potential explanatory causal factors for individual differences in chess skill. For example, pioneering work from information-processing models, perception, and memory judged the appropriateness of abilities involving abstract relations for chess, far beyond the organization of chess information into chunks. Furthermore, although there appeared to be a lack of evidence regarding the superiority of stronger players on basic intellectual factors, exceptional abilities were argued as indispensable at a World Chess Championship level (Chase & Simon, 1973; Simon & Chase, 1973). The interest in intellectual factors as studied in the chess domain is further exemplified by more recent meta-analytic reviews, which have summarized the available empirical evidence to date (Burgoyne et al., 2016; Sala, Burgoyne et al., 2017). On the other hand, studies involving the personality of chess players together with the interests implied in becoming involved in a long and demanding chess learning process are much more scarce (Grabner, Stern, & Neubauer, 2007).

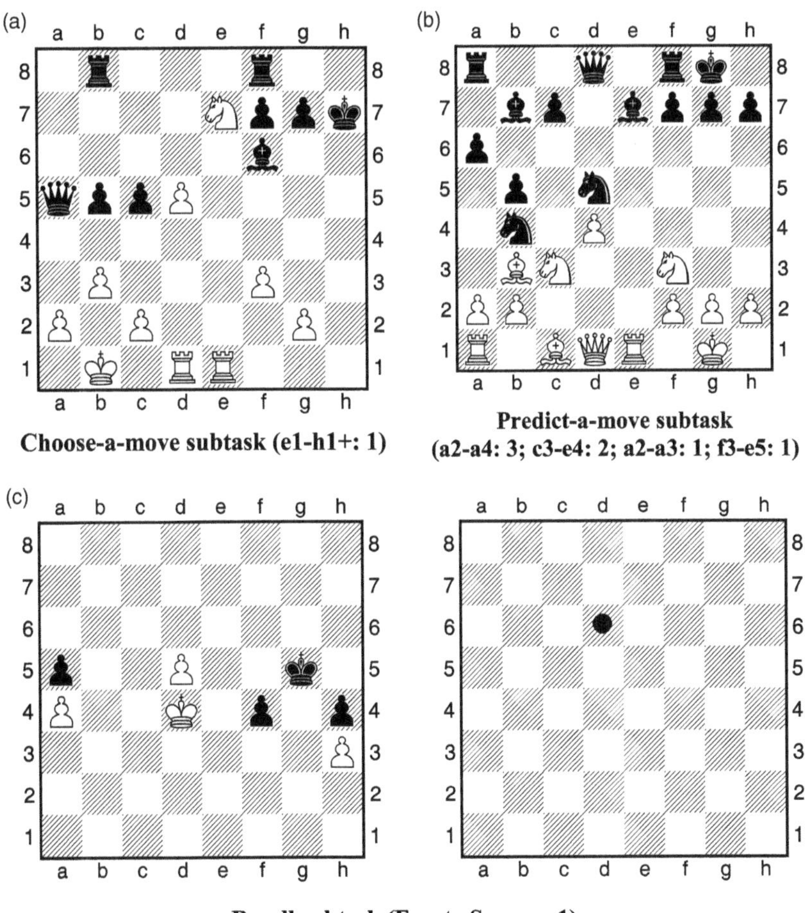


**Figure 4.6** Sample items from the Amsterdam Chess Test in the choose-a-move, predict-a-move, and recall subtasks (van der Maas & Wagenmakers, 2005)

The PPIK theory is a useful conceptual approach for studying the association of several traits with individual differences in chess skill. The PPIK framework encompasses four broad components that are helpful to organize and examine the structure and development of the psychological traits presumed to have a bearing on chess skill (Ackerman, 2007). Is there any psychometric test to evaluate individual differences in traits embedded within the PPIK components, however? Yes, there is. The Amsterdam Chess Test (ACT) is an empirical tool that allows for the evaluation of traits from some of the four broad components of the PPIK theory (van der Maas & Wagenmakers, 2005). The development of the ACT is probably the most comprehensive psychometric study undertaken about individual differences in chess. Using data from a considerable sample of chess players during a chess tournament in the Netherlands (N = 259), this study describes the development of the ACT, a comprehensive computerized instrument aimed at calibrating chess skill. The development of the ACT was motivated by four main issues: the lack of a psychometrically sound measurement device to evaluate chess skill; the in-depth evaluation of psychometric techniques within an intellectually demanding yet systematic domain; the examination of the nature of chess proficiency; and for pragmatic reasons related to the initial evaluation of chess skill.

The ACT comprises five subtasks delivered through a computer: a choose-a-move subtask (two tests, A and B), a chess motivation questionnaire, a predict-a-move task, a verbal chess knowledge questionnaire, and a recall subtask (Figure 4.6). Each of the two tests from the choose-a-move subtask comprise forty typical chess problems, including twenty tactical items, ten positional items, and ten endgame items, all of them to be completed within thirty seconds. The sample tactical item shows that the correct answer is the move e1-h1+, which scores one point, whereas any other answer scores zero points. The motivation questionnaire comprises thirty items measuring three motivational traits: positive fear of failure, negative fear of failure, and desire to win. There are ten seconds to complete each of these items, which are answered on a five-point 'disagree'–'agree' scale. The predict-a-move subtask comprises forty-two items corresponding to the course of a real chess game (Liss versus Hector, Copenhagen, 1996), with thirty seconds allotted to each item. The participant is asked to suggest the best continuation for the white pieces, with a scale from zero to five points. For the sample item, the suggested moves scored either three (a2-a4), two (c3-e4), or one (a2-a3 or f3-e5) points, while any other move scored zero points. The verbal chess knowledge questionnaire comprises eighteen four-choice items about the opening, positional, endgame, and visual imagery aspects of the game. There are fifteen seconds to complete each of these items. The recall subtask consists of eighteen items. An initial chess diagram is presented for ten seconds (left diagram in Figure 4.6 c), then a blank screen follows for two seconds, and then an empty chessboard with one square marked by a circle (square d6 in the right diagram in Figure 4.6). The participant has to identify within ten seconds the piece in the initial diagram that was occupying the square marked with the circle in the empty diagram. The correct answer for the item sample is 'empty square'. Higher scores in the five subtasks from the ACT indicate a higher level of the respective evaluated traits.

In the study describing the development and application of the ACT (van der Maas & Wagenmakers, 2005), the subtasks of this instrument were highly predictive of the variability in chess skill as measured by the Elo rating (75%), together with sex, age, and the number of chess games played the season prior to the data collection. Unfortunately, even though the ACT provides a comprehensive evaluation of chess skill independently of the commonly used Elo rating, it has been rarely used in applied research about individual differences in chess. As suggested by the authors of the ACT, the test could be used to estimate individual differences in chess skill independently of the Elo chess rating.

## 4.3 Heredity versus Environment

What are the reasons and motives leading individuals to play chess? More importantly, why do only a few individuals achieve a remarkable performance in the game? When somebody reaches a grandmaster chess level, is such an accomplishment a matter of innate predisposition, or does it depend on a long period devoted to intensive and specific chess training? These are recurrent issues that frequently loom large in differential psychology and chess research. Whether behaviour depends on the influence of genes (nature) or on the influence of the environment (nurture) is arguably the most cogent debate in social sciences. The problem has elicited a considerable deal of research effort to date, but the controversy might be already over. It has been claimed that the distinction between nature and culture as two differentiated entities affecting behaviour should be dismissed, because organisms inherit regularities from both genes and the environment (Buss, 2001; Pinker, 2002; Tooby & Cosmides, 1990, 2005; Tooby, Cosmides, & Barrett, 2003).

The influence and mechanisms of both systems in determining behaviour is one of the most groundbreaking topics in the field (Bleidorn et al., 2010; Kanai & Rees, 2011; Plomin et al., 2014). How genes and environment combine to shape human behaviour was thought to articulate around six main lines of research (Anastasi, 1958):

1. the associations between physiological variables and behavioural variations;
2. the role of prenatal physiological factors in later development;

3. the influence of early experiences on later behaviour;
4. the cultural differences in child-rearing concerning development;
5. the mechanisms of somatic and psychological relationships; and
6. the development of twins and their social environment.

More modern methods and technologies have served to implement these lines of research, such as genome-wide sequencing technologies (Mardis, 2009), knowledge management systems (Ashburner et al., 2000; Blanch, García, Planes, et al., 2017), and statistical and conceptual developments (Franic et al., 2013; Günther, Wawro, & Bammann, 2009; Hodgins-Davis & Townsend, 2009).

Human behavioural genetics is the scientific field addressing how the genotype, the inherited component encoded in the genes, interacts with the environment to give rise to complex traits or phenotypes (Plomin & Rende, 1991). The main aim in human behavioural genetics is to discover how the variation in traits emerges from the complex interaction between inherited components and environmental influences. Three main laws in human behavioural genetics are that (1) all human traits are heritable, (2) the effect of being raised in the same family is smaller than the effect of genes, and (3) a substantial portion of the variation in complex human behavioural traits is unaccounted for by the effects of genes or families (Turkheimer, 2000). From this viewpoint, genes and environments embed into a development system, eventually producing individual differences in behaviour. With additive and independent genetic and environmental effects, a given trait ($x$) in a population would arise from the combination of genetic (G) and environmental (E) factors, as shown in Equation 3.1, where $e$ is a quantity due to trait measurement unreliability. Within a given population, and as shown in Equation 3.2, the variability in the trait (variance ~ standard deviation) is due to the variability in genetic ($var$(G)) factors, the variability in environmental ($var$(E)) factors, and the correlation of genetic and environmental factors (2$Cov$(G,E) (Hunt, 2011).

$$x = G + E + e \tag{3.1}$$

$$var(x) = var(G) + var(E) + 2Cov(G, E) \tag{3.2}$$

The extent to which the variability in the trait is heritable in the observed population depends on the proportion of phenotypic variance due to genotypic variance across individuals in the population (Plomin, DeFries, & Loehlin, 1977; Turkheimer, 1991, 1998). This is typically characterized with the heritability coefficient ($h^2$), as shown in Equation 3.3, keeping in mind that $h^2$ is applicable only to populations, not to individuals.

$$h^2 = var(G)/var(x) \tag{3.3}$$

Human behavioural genetics draws on data from three kinds of studies. Family studies compare members of the same family who share genetic and environmental components. Twin studies compare monozygotic (MZ) and dizygotic (DZ) twins who share 100 and 50% of genetic components, respectively. Adoption studies include individuals who share environmental factors only, together with individuals who share both genetic and environmental factors. In general, the extent to which individuals of varying genetic relatedness are more similar in a given trait supports a greater influence of either genetic or environmental factors. For instance, and in a twin study, a higher correlation between MZ than between DZ twins in a given trait would support a higher genetic predisposition regarding that trait (Plomin et al., 1997).

The available empirical evidence from human behavioural genetics supports the view that most, if not all, individual differences in behaviour are partially heritable. Genes affect complex behaviour, but so do shared and non-shared environments. The shared environment comprises the influences that make family members similar to one another. The non-shared environment comprises the effects that make family members different from one another, such as the individual experiences of children with their peers, and the facets of parenting that are not shared in the same family (Plomin & Daniels, 1987; Turkheimer & Waldron, 2000). For example, both genetic and environmental factors might have an equivalent weight in determining group differences in intelligence (Rushton et al., 2007). Thus, the environment is also a fundamental aspect in shaping development and complex behaviour when interacting with genes, because it contributes to changing subsequent behaviour and the expression of genes in a reciprocal cycle (Plomin & Petrill, 1997; Plomin & Rende, 1991; Turkheimer, 1991, 1998, 2000).

The combination of variation in genotypes and in environments in producing individual differences in development was addressed in the early 1980s, suggesting three kinds of genotype → environment effects: passive, evocative, and active (Scarr & McCartney, 1983). The passive kind takes place in genetically related families, with parents providing genes and environments for their offspring. The evocative kind contemplates those instances whereby different genotypes evoke different responses from the environment. The active kind consists in the choices made by individuals from different environments, which tend to match their motivational and intellectual genetic endowment. Two compelling additional propositions of this model are that, during development from infancy to adolescence, the effect of the passive kind decreases while the effect of the active kind increases. A later bioecological model advanced three testable hypotheses pivoting on the concept of proximal processes as the core organism–environment interaction mechanisms directly implicated in development. Proximal processes comprise parent–child and child–child activities, group or solitary play, reading, learning new skills, problem solving, performing complex tasks, and acquiring new knowledge. In this view, weaker proximal processes impair genetic potentials, whereas stronger proximal processes enhance genetic potentials. More specifically, stronger proximal processes would allow for the accomplishment of important developmental outcomes, such as coping under stress, acquiring knowledge, and establishing rewarding relationships (Bronfenbrenner & Ceci, 1994).

1. The heritability ($h^2$) will be higher with stronger proximal processes, and lower with weaker proximal processes.
2. Strong proximal processes will contribute more to developmental competence in advantaged and stable environments. Strong proximal processes will buffer more developmental dysfunction in disadvantaged and disorganized environments.
3. Strong proximal processes will contribute more to developmental competence for those individuals in disadvantaged and disorganized environments when they are exposed extensively to circumstances rich in developmental resources.

Genetic and environmental underpinnings of personality and intelligence have stimulated a great deal of research (Caspi et al., 2005; Plomin, Owen, & McGuffin, 1994). Meta-analytic findings suggest that 40% of the variation in personality is due to genetic influences (Vukasovic & Bratko, 2015). Moreover, genetic and non-shared environmental factors account for personality changes that tend to stabilize at later ages (Hopwood et al., 2011). For instance, findings with over 21,000 sibling pairs indicate more stable genetic and environmental effects on personality with ageing, and with additional increments of environmental effects on phenotypic stability into adulthood (Briley & Tucker-Drob, 2014). Findings supportive of the universality of human personality advocate in addition for a consistent biological basis of personality (Yamagata et al., 2006). The association of personality traits with genetic polymorphisms (5HTT LPR, DRD4 c4t, DRD4 length, DRD2 A1/A2, DRD3 A1/A2) have rendered non-significant findings after controlling for allele frequencies and unpublished data, however (Munafó et al., 2003).

As with personality, there seems to be a weak association between general intelligence (*g*) and single-nucleotide polymorphisms in several genes (DTNBP1, CTSD, DRD2, ANKK1, CHRM2, SSADH, COMT, BDNF, CHRNA4, DISC1, APOE, and SNAP25). This finding has been attributed to low statistical power linked to limited sample sizes (Chabris et al., 2012). In contrast, general intelligence might have a consistent polygenic structure underlying multiple tiny effects from different gene loci and their combinations (Chabris et al., 2013). This polygenic approach to the genetic influence of complex traits had, in fact, already been advanced in earlier studies about the genetic basis of human behaviour (Bouchard & McGue, 1981; Plomin et al., 1994), and is actually being implemented at present with modern genome-wide association scan applications (McCrae et al., 2010). Several twin and adoption studies point to a considerable amount of variability in intelligence being accounted for by genetic factors (Johnson et al., 2007; Plomin & Spinath, 2004; Posthuma, de Geus, & Boomsma, 2001). Moreover, studies with a focus on the development of intelligence highlight three main findings: genetic influences increase throughout the lifespan; the importance of shared environment in childhood and of non-shared environment beyond adolescence; and stronger genetic influences in more favourable social and economic conditions (Plomin & Petrill, 1997; Plomin & Spinath, 2004; Tucker-Drob & Briley, 2014; Tucker-Drob, Briley, & Harden, 2013).

Behavioural genetics has also focused on socially relevant outcomes produced in important contexts such as family, work, and education. In examining the affective climate in over 600 families, genetic factors accounted for 37%, whereas shared environmental factors accounted for 19% of the variability in the expression of negativity such as anger or hostility, as opposed to the expression of positivity such as warmth or affection (Rasbash et al., 2011). In the work realm there are several explanatory mechanisms involving genetic effects on work attitudes and leadership (Ilies, Arvey, & Bouchard, 2006), positive and negative affectivity associated with job satisfaction (Ilies & Judge, 2003), and coping with work demands (Maas & Spinath, 2012).

Because personality and intelligence have been shown to be highly heritable, and intelligence is a very robust predictor of educational achievement (Aluja & Blanch, 2004; Blanch & Aluja, 2013; Deary et al., 2007; Poropat, 2009), educational accomplishment has attracted considerable attention from human behavioural genetics. For example, the CoSMoS German study with ninety-seven MZ twins and 183 DZ twins corroborated a meaningful heritability of cognitive ability, self-perceived abilities, and academic achievement (Gottschling et al., 2012). Moreover, data from a sample in the Minnesota Twin Family Study (MTFS) suggested that shared and non-shared environmental influences on educational achievement were stronger at low intelligence quotient (IQ) values, decreasing with IQ increments (Bouchard & McGue, 1981). In contrast, genetic influences on educational achievement were reported as weaker at lower IQ, while increasing at higher IQ (Johnson, Deary, & Iacono, 2009). Further data from the Twins Early Development study (TEDS) confirmed that intelligence accounted for the higher proportion in the heritability of academic achievement. Individual differences in other factors, however, such as self-efficacy, personality, and behavioural problems, accounted for nearly as much heritability of academic achievement as intelligence did, supporting additional influences of other traits with a robust genetic basis (Krapohla et al., 2014). Genome-wide association studies (GWASs) with over 126,000 individuals have identified genetic variants associated with educational achievement (Rietveld et al., 2013), while linking polygenic scores (PSs) with educational attainment scores and relevant outcomes in life development such as adult economic outcomes, the acquisition of speech and reading skills, geographic mobility, mate choice, and financial planning (Belsky et al., 2016).

As far as it is known, there are no behavioural genetics studies with chess players. Nonetheless, the controversy of nature versus nurture has also emerged to some extent in chess studies. For instance, a comparison of sixteen novice, eight intermediate, and eight expert chess players suggested a greater advantage for experts in perceptual encoding, which, it is argued, depends on chess experience and knowledge rather than on an innate predisposition involving a perceptual or memory superiority (Reingold et al., 2001). Similarly, another comparison of forty children with forty adults (twenty novice and twenty expert chess players in children and adults alike) supported the view that specific chess knowledge impinged clearly on memory performance. The advantage of children was argued to be associated with innate memory abilities important for chess, however, in contrast to adults, who would be more dependent on learning and accumulated knowledge from experience (Schneider et al., 1993).

The nature versus nurture controversy is of considerable theoretical importance for the development of chess expertise, particularly when considering whether chess skill depends on individual dispositions or on the effects of practice. For example, some authors suggest practice as the crucial element for the development of chess expertise (Ericsson & Charness, 1994; Ericsson, Krampe, & Tesch-Römer, 1993). In contrast, other authors argue that practice is a necessary but not a sufficient condition for achieving an expert chess level, suggesting that only a modest part of the variability in chess skill is due to the effects of practice (Campitelli & Gobet, 2011; Hambrick et al., 2014). In any event, comprehensive theoretical accounts about the development of expertise, including chess, have attempted to integrate multiple genetic and environmental factors (Ericsson et al., 2006; Gobet, 2016; Simonton, 2014a, 2014b; Ullén, Hambrick, & Mosing, 2016). This topic is addressed in greater depth in a later chapter, on expertise in chess (Chapter 8).

An initial place to look for individual differences in chess skill is the psychophysiology and brain functioning of chess players, however; a fascinating theme that is addressed in depth in the next chapter.

# 5

# Psychophysiology and Brain Functioning

Chess is a natural domain appropriate to study issues related to individual differences in psychophysiology and brain functioning. It has simple and universal rules, it implies higher-order cognition and domain-specific knowledge, and the Elo rating is a valid and reliable quantitative measure of chess performance (Charness, 1992; Gobet, 1998; Gobet & Charness, 2006; Simon & Chase, 1973; van der Maas & Wagenmakers, 2005). Chess is analogous to other sports regarding biochemical, physiological, neuronal, and psychological aspects. Because of its main focus on intellectual performance, however, chess is unique compared to other sports of a more physical nature. The energy required by neuronal activation is delivered by oxygen and glucose through the vascular system. Higher neuronal activity implies higher demand for oxygen, which increases in turn the brain blood volume and flow (Golf, 2015a).

An increasing amount of studies address psychophysiological and brain activity parameters during chess playing. For instance, there is a recent multimodal magnetic resonance imaging (MRI) dataset from twenty-nine professional chess players available to interested researchers, which includes phenotype data such as age, sex, education, and intelligence measures (Li et al., 2015). Moreover, comprehensive reviews highlight that brain anatomy has robust links with inter-individual differences in motor behaviour and learning, perception, cognition, intelligence, and personality (Kanai & Rees, 2011). Because chess is a highly demanding intellectual activity involving several of these attributes, the psychophysiology and brains of chess players constitute an attractive object for study in a controlled domain.

This chapter comprises two main areas of enquiry. The first focuses on the psychophysiology of chess players by reviewing research works about the heart rate (HR), respiratory variables, hormones, and the issue of doping in chess. This area encompasses a smaller amount of studies comprising physiological functioning events other than brain functioning. The second area of enquiry focuses on the brain activity of chess players with techniques such as electroencephalography, functional magnetic resonance imaging, magnetic resonance imaging, magnetoencephalography (MEG), positron emission tomography, and single-photon emission computerized tomography (SPECT).

## 5.1 Psychophysiology and Chess

Most studies addressing psychophysiological parameters in chess have been framed in terms of the human stress response model. This model proposes that activation of the sympathetic nervous system, a parasympathetic withdrawal, and heightened activity of the hypothalamic–pituitary–adrenal axis contribute to physiological and mental improvements to meet environmental demands (Sapolsky, 1996; Selye, 1975). The chess domain has been used to evaluate such psychophysiological stress mechanisms under a stressful condition involving a high cognitive workload. One of the most meaningful physiological markers in stress processes is the heart rate (HR), which measures the heart's contractions or beats per minute (bpm), with a normal range of between 60 and 100 bpm in a resting state, and with abnormalities in this range being associated with a higher risk of cardiovascular disease (Fox et al., 2008). In chess, HR is frequently associated with several playing circumstances, such as committing an irrecoverable mistake that usually leads to a defeat (i.e., a blunder), the evaluation of a difficult move, a piece sacrifice, a hazardous defence, or a winning move (Golf, 2015a).

A pioneering study examined parameters such as psychological maximum stress, autonomic excitability, circulatory conditions, and the long-term metabolic load during an eighteen-day tournament with fourteen chess players (Pfleger et al., 1980). Over 70% of the participants showed increased cholesterol levels in blood throughout the chess tournament, whereas their autonomic excitability and circulatory parameters were deemed to be fully comparable to those reported in other sports. Heart rate variability (HRV) has also received a certain degree of attention. This physiological parameter indicates the time between two heartbeats. A low HRV is associated with increased mortality, the incidence of cardiac events, and a heightened risk of sudden cardiac death (Achten & Jeukendrup, 2003). For example, cardiac events such as high-frequency heart rate variability (HF-HRV) relate to a psychometric measurement of helplessness/hopelessness, as the opposite trait to optimism/control, applying to nine active players above 2300 Elo rating points (Schwarz et al., 2003). Increased hopelessness associated with a reduced HF-HRV, suggesting a consistent negative correlation between negative mood states with autonomic nervous system disruptions, and with observable cardiac events. In addition, another intriguing finding is that, despite the very broad and basic nature of HR as a psychophysiological marker, it related to cognitive events during twenty-five chess games played by nine chess players within a range of about 2021 to 2216 Elo rating points. Furthermore, framing and processing the attainment of a specific goal through planned action and analysing variations and opponent's blundering predicted meaningful increments in the incidence of HRV, with a 75% chance of a reliable detection of HRV changes for the analyses of chess variations (Leone et al., 2012). It has also been suggested that HRV decreased during a rapid game between an elite human chess player with an Elo rating of 2550 points and a machine (Fuentes et al., 2018). These meaningful connections between HRV and cognitive processes run parallel to experimental outcomes in a sustained attention task, whereby there was higher HRV in a group with high physical fitness compared with one with low physical fitness from the general population (Luque-Casado et al., 2013).

In addition to HRV, there is evidence on the association of the stress caused by chess playing with respiratory variables such as ventilatory flow (VF), tidal volume (Vt), breath frequency (bF), $O_2$ consumption ($VO_2$), $CO_2$ production ($VCO_2$), and respiratory exchange ratio (RER), among others (Troubat et al., 2009). Measures of these parameters were obtained from twenty chess players with between 1250 and 2170 Elo rating points when playing a single one-hour game against computer software set at 100 Elo points above each human's player respective Elo rating. All measures were obtained before and during the game. There was a meaningful increment in HR throughout the game, supporting the sympathetic nervous system stimulation view. Furthermore, the HR was higher at the start of the game, then decreased in the latter stages, suggesting that energy expenditure switched from carbohydrate to lipid oxidation. This body of evidence is interesting in its own right for the field of human stress, as it provides sound findings from an ecologically valid and well-controlled domain such as chess. On the other hand, however, it is somewhat surprising that none of the aforementioned studies has addressed the impact of individual differences in chess skill as related to the obtained psychophysiological findings. Moreover, there is a remarkable dearth of studies addressing the role of certain hormones on chess performance. For example, testosterone has been associated with chess tournament outcomes. Winners of chess tournaments showed higher concentrations of testosterone than losers did, even before actually playing the games. Heightened testosterone appeared to reinforce a dominance pattern of behaviour and aided the winning of these particular chess contests (Mazur, Booth, & Dabbs, 1992).

The sporting nature of chess includes a polemic issue with many legal and medical implications: doping in chess playing. Pharmacological products and a variety of new drugs appear to improve cognitive functioning in attention, memory, learning, and executive functions (Fond et al., 2015; Repantis et al., 2010). Several substances, such as stimulants, narcotics, anabolic agents, diuretics, and peptide and glycoprotein hormones and their analogues, are prohibited in a variety of sports, such as athletics, cycling, soccer, and swimming. In contrast to physical performance, however, doping effects appear to be far more complex regarding cognitive performance. In the paradigmatic domain of chess, the effect of potential pharmacological enhancers is, apparently, unwarranted (Mihailov & Savulescu, 2018).

Nevertheless, the World Chess Federation (FIDE) adopted a drug policy in 1999, and oversees doping controls during major chess tournaments. A recent review has suggested that an improvement in brain performance in chess playing may be achieved with increments in the $O_2$ supply by therapy with erythropoietin (EPO), increases of body glycogen by therapy with insulin, or with mental stimulation by caffeine. In addition, androgenic-anabolic substances, amphetamines, nicotine, and cocaine appear to exert innocuous effects on the quality of chess playing (Golf, 2015b). On the other hand, some steroids and hormones have positive effects on cognition only when present in natural concentrations. Besides, pharmaceutical preparations show positive effects only at low baseline cognition, whereas these hormones present negative effects on mental cognition at elevated concentrations. For example, a randomized controlled trial with thirty-nine chess players who were administered modafinil, methylphenidate, caffeine, or a placebo suggests that these substances do improve decision-making, albeit performance worsened notably with time constraints (Franke et al., 2017).

## 5.2 Brain Basics

The brain is an intriguing organ. Brain anatomy and brain structures and functioning are certainly complex, with between 150 and 200 highly interconnected cortical areas. In addition, there is a high degree of inter-individual variability in these structures and functions, and several notable brain interhemispheric differences. A number of human brain atlases with microstructure data and cortical segregation tools have been developed to explore what is probably one of the fundamental frontiers of scientific knowledge (Amunts & Zilles, 2015). In sharp contrast to other species, the cerebral cortex is the most recent brain structure to have evolved in humans, underlying the anatomical basis of cognitive abilities (Rakic, 2009). The cerebral cortex governs several properties and functions closely related to chess performance, such as perception, memory, and higher-level abstract skills, such as concentration, reasoning, and thinking.

The brain is divided into four main lobes in the right and left brain hemispheres: frontal, parietal, temporal, and occipital (Clark, Boutros, & Mendez, 2010; Hill & Schneider, 2006). The frontal lobes control motor skills, including voluntary movement and speech, and intellectual and behavioural functions particularly important for chess, such as problem solving and judgement. The parietal lobes involve sensory awareness, symbolic communication, and abstract reasoning. The occipital lobes operate the visual-processing system. The temporal lobes are implicated in visual memory and in the recognition of objects and faces, and in verbal memory for the use of language. These are only the main basic functions of the brain lobes, as there is a high degree of interactions within and across these main brain regions. For example, the neural reuse theory provides an alternative approach concerning the organization of brain modules and the localization of cognitive functions (Anderson, 2010).

Brain white matter is a densely packed assembly of myelinated projections of neurons, whereas brain grey matter is composed of the cell bodies of neurons. White matter seals the four lobes together and provides the connectivity among the four regions into specific networks designed to execute a variety of mental operations. In addition, ridges (*gyri*) and grooves (*sulci*) characterize the surface of the cerebral cortex. Deeper *sulci* are also termed fissures, with the most remarkable one being the inter-hemispheric fissure, which separates the two brain hemispheres. The Figure 5.1 (a) shows the four main lobes of the human cerebral cortex: frontal, parietal, temporal, and occipital.

The study and direct *in vivo* observation of the brain has experienced a considerable expansion during the last few decades because of the growing availability of sophisticated techniques. These permit the study of what is going on in the brain in a diversity of circumstances (Volkow, Rosen, & Farde, 1997),

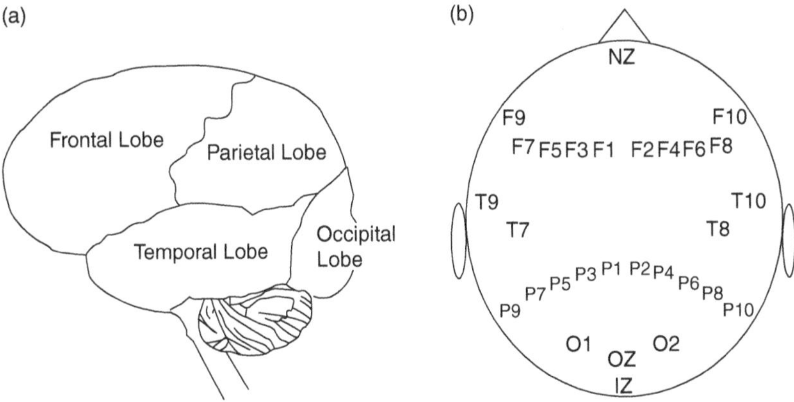


**Figure 5.1** Diagram (a): the four main lobes of the human cerebral cortex: frontal, parietal, temporal, and occipital. The main functions of the cerebral cortex are (frontal lobes): motor skills, voluntary movement, speech, problem solving and judgement; (parietal lobes): sensory awareness, symbolic communication, and abstract reasoning; (occipital lobes): visual processing; (temporal lobes): visual memory, recognition of objects and faces, and verbal memory for the use of language (reproduced with permission from the American Psychological Association). Diagram (b): the 10–20 system for the recording of electroencephalograms (EEGs) in humans, showing the reference electrodes nasion (NZ) and inion (IZ), and electrodes corresponding to the cerebral cortex lobes, frontal (F), temporal (T), parietal (P), and occipital (O); the amount of electrodes can vary depending on the research aims and kind of equipment

including those involved in chess playing and chess expert performance and acquisition. The study of the brain functioning of chess players can be classified into two main research streams. The first stream deals with electroencephalograph. EEG was the very first method used to analyse brain activity, and it probably remains the cheapest and most used nowadays. The second stream comprises other brain-imaging techniques, and also incorporates some of the general findings obtained with EEG. In general, brain activity in chess players has been observed to vary in accordance with the brain region being analysed together with the level of chess expertise.

## 5.3 Electroencephalography (EEG)

EEG is a widely used technique applied to the measurement, recording, and representation of the brain's activity. It is an electrical biosignal produced by aggregated electric potential differences elicited from neuronal activity. This signal is obtained with computer hardware and software, which includes the standard 10–20 system recording cap placed on the scalp of the subject (see diagram (b) in Figure 5.1). Each electrode is designated by letters and numbers that identify the position in the names of the main brain lobes (Acharya et al., 2016). The EEG amplitude (voltage) varies between 1 and 100 μV in a normal healthy adult depending on the stimulation and the subject's mental state. Different frequency bands characterize EEG in different brain states. Higher frequencies imply a higher level of brain activity. For example, the delta band (from 0.1 to 3 Hz) relates to an unconscious state such as sleep, the alpha band (from 8 to 12 Hz) relates to a relaxed but conscious state, and the gamma band (from 30 to 100 Hz) relates to motor functions and a higher mental activity (Carretié, 2001).

EEG has a high temporal resolution and comprises studies in three broad areas: spontaneous brain activity, single-neuron bioelectric events, and event-related potentials (ERPs). The ERP application is the most commonly used in experimental psychology studies, with the main focus being placed on mental processes such as perception, attention, language processing, memory, and decision-making. The ERP application aims to detect the brain's electrical activity in response to diverse physical stimuli when associated with mental activity or in preparation for specific actions. For example, two classical ERP components are the N200 and P300 negative deviations (N), appearing between 185 and 325 milliseconds after the onset of an auditory or visual stimulus, whereas positive deviations appear between 300 and 400 milliseconds afterwards. Both components have been consistently linked with perceptual and attention cognitive processes (Patel & Azzam, 2005). Individual differences in relevant constructs for chess performance relate to EEG, such as semantic memory and the ability to learn new material (Doppelmayr et al., 2002), general creativity (Dietrich & Kanso,

2010) and verbal creativity (Fink & Neubauer, 2006), and mental arithmetic tasks (Duru & Assem, 2018; Garach et al., 2015).

EEG is a complex technique that is highly demanding, both technically and computationally (Kenemans, 2013; Weiergräber et al., 2016). Therefore, some studies with chess players focus on the evaluation of EEG data analysis techniques, such as the Higuchi fractal dimension, with non-linear methods (Stepien, Klonowski, & Suvorov, 2015). Research more substantially linked to the brain activity of chess players has addressed more specific hypothesis. For instance, the EEG theta Fz/alpha Pz ratio was used as an estimation of the brain load at pre-game, chess game, and post-game stages with a 2550 Elo rating points elite chess player (Fuentes et al., 2018). As expected, the brain load was higher with the greater demand of information processing during the chess game stage. A key finding, however, was the higher activation in the prefrontal cortex at the pre-game stage. The brain load at the next game stage was anticipated analogously to the association of testosterone levels with winning a chess contest before it actually takes place (Mazur et al., 1992). This single-subject study design precluded further explanations as to how individual differences in chess expertise could modulate this behaviour, however.

Individual differences between expert and novice chess players during EEG recordings have also been reported by using recognition tasks of chess information. Eight high-proficiency chess players compared with eight low-proficiency chess players underlined significant differences when eye-tracking chessboards comprising five differentiated conditions: piece positioning, check situations, checkmate situations, checkmate in one move, and capturing pieces (Silva-Junior et al., 2018). High-proficiency players fixed their eyes on relevant parts of the position with more EEG activation in brain areas related to planning and decision-making, such as the prefrontal cortex. In contrast, low-proficiency players fixed their eyes on larger spaces, implying the processing of a greater amount of information, with more brain activity in areas related to the primary processing of vision and eye control, such as the occipital and parietal lobes. These findings are somewhat aligned with the idea that prefrontal cortex activation might be particularly relevant for more proficient individuals when working with associative relationships, and planning (Clark et al., 2010).

Another study exposed eleven expert and eleven novice chess players to four kinds of recognition tasks: whether there was a white king on the board, whether or not the black king was in check, whether or not the black king was checkmated, and whether or not a specified chess piece could checkmate the black king in the next move (Volke et al., 2002). The study examined the degree of coherence – i.e., the extent of the relationships between different cortical areas in resolving the given tasks. Experts showed well-mastered task resolution and automatic performance (i.e., high coherence), whereas novices showed poorly mastered task resolution (i.e., low coherence). Recognition stimuli have also been used to examine individual differences between chess experts and novices in the N200 and P300 EEG components (Wright et al., 2013). Fourteen expert chess players and fourteen novice chess players attempted to recognize a check position, the presence of a black knight, a non-check position, and the absence of a black knight. Expert players had a more pronounced N200 in the check tasks and a larger P300 in the knight tasks compared to novices, suggesting a better harmonization of perceptual inputs into memory – i.e., superior pattern recognition and chunk retrieval by experts over novices. These findings support consistent individual differences between experts and novices in information processing, providing additional support for the template theory of expert memory.

Further research has analysed a compelling hypothesis of intelligence with data from chess players. The neural efficiency hypothesis (NEH) maintains that individual differences in intelligence relate to individual differences regarding brain activation during cognitive tasks. More specifically, the NEH predicts that, when confronted with a relatively demanding cognitive task, people who are more intelligent tend to display lower brain activation, whereas people who are less intelligent tend to display higher brain activation (Neubauer & Fink, 2009a). Alpha band event-related desynchronization (ERD) is a valid EEG method to characterize the degree and topographical distribution of cortical activation in addressing the NEH. The ERD posits that the alpha EEG band within 7.5 and 12.5 Hz changes during cognitive activity. Decreases in ERD are taken as increased neural excitability and cortical activity, whereas increases in ERD are taken as decreased neural excitability and cortical inactivity. This paradigm was applied to fifty-five people from the general population in experimental tasks involving short-term and working memory, and attention switching in the central executive task (Grabner et al., 2004). Overall, the NEH was supported to a greater degree for fluid as opposed to crystallized intelligence, for males compared with females, and regarding the interaction of type of task with brain area by sex. There were positive associations of fluid intelligence with ERD for males in the frontal lobe when performing the central executive task.

The ERD method was also applied to another sizable sample of forty-seven chess players involving mental speed, memory, and reasoning tasks (Grabner, Neubauer, & Stern, 2006). This experiment compared in addition the association of both intelligence and chess skill with the elicited cortical activation pattern. These latter ERD findings with the chess sample were surprisingly similar to those findings obtained from the study with the non-chess sample. In accordance with the NEH, individuals who scored higher in cognitive ability had lower brain activity than individuals who scored lower in cognitive ability throughout the three experimental tasks. Furthermore, the activation of the prefrontal brain cortex was more efficient in individuals scoring higher in cognitive ability. When taking into account the level of chess skill and concerning the short-term memory and reasoning tasks, chess players with a higher level of chess skill had lower activation of the frontal cortices, albeit higher activation of the parietal cortices. In contrast, more skilled players also displayed higher cortical activation than less skilled players during the memory task, in virtually all brain areas. This finding was attributed to the manipulation of verbal-semantic information during memory tasks, which might indeed reflect higher efficiency in mobilizing prior knowledge resources. Overall, the main picture of the studies addressing the NEH underline the remarkable individual differences in brain activation, but also in intelligence and chess skill level, with brighter players displaying higher efficiency in brain functioning.

## 5.4 Overview of Brain-Imaging Studies

Brain-functioning study techniques tend to be costly, require accurate execution, and impose severe and intensive computational requirements. Nonetheless, the number of brain-imaging studies with chess players and experts in several fields has experienced progressive growth in the past few years. A recent comprehensive investigation scrutinizes individual differences between experts and novices from a neuroimaging approach, in domains such as chess, mental calculation, and perceptual and motor expertise in several occupations (Bilalić, 2017). Table 5.1 shows an overview of most of the twenty-six brain-imaging studies performed with chess players, summarizing the characteristics of the sample, the kinds of tasks performed during brain scanning, the brain-imaging method, the brain areas and structures being analysed, and the main basic findings. The sample size ranged from a single chess player (Fuentes et al., 2018) to forty-seven chess players (Grabner et al., 2006), with a mean sample size of sixteen chess players (*Sd* = 11). Over 90% of the participants in these studies were males. Furthermore, several studies compared a group of experts against a group of novice chess players, whereas only nine studies (36%) did not conduct such kind of comparison. Most studies considered the Elo rating as the index by which to separate experts from novices, even though some studies applied an independent test to either classify or verify the level of chess skill of the participants (Silva-Junior et al., 2018; Volke et al., 2002; Wright et al., 2013).

All the studies used visual stimuli during chess playing, chess problem solving, and memory or recognition tasks. One study used only verbal stimuli within a blindfold chess paradigm (Saariluoma et al., 2004). There were five main kinds of behavioural tasks or paradigms applied to each study: playing a chess game against a computer, solving chess positions, recognizing chess positions, memorizing chess positions, and completing self-report questionnaires. In some studies, two or more of these tasks were combined into a single experimental paradigm. Moreover, a few studies also included non-chess aspects, such as face, scene and object recognition, or theory of mind and empathizing tasks (Bilalić, Langner, et al., 2011; Krawczyk et al., 2011; Powell

Table 5.1 *Studies in brain functioning in chess players (EEG = electroencephalography; MEG = magnetoencephalography; fMRI = functional magnetic resonance imaging; PET = positron emission tomography; SPECT = single-photon emission computerized tomography). In the N column, 'M' denotes that all participants were males, and 'F' denotes the number of females*

| Study | N | Tasks/paradigm | Method | Main brain areas/ structures | Main conclusion(s) |
|---|---|---|---|---|---|
| (Amidzic, Riehle, & Elbert, 2006) | 10 experts (2400 ~ 2600) 10 novices (> 1700) (M) | Chess game against a computer | MEG (gamma-band activity, 20 – 40 Hz) | Medial temporal lobes; frontal and parietal cortex; hippocampus | There are more frequent focal gamma bursts in deeper structures of the medial temporal lobes of amateur players. There are more frequent gamma bursts in the frontal lobes of chess grandmasters. |
| (Amidzic et al., 2001) | 10 experts (2400 ~ 2600) 10 novices (> 1700) (M) | Chess game against a computer | MEG (gamma-band activity, 20 – 40 Hz) | Medial temporal lobe; frontal and parietal cortices | Amateur players show higher activity in the medial temporal lobe. Grandmasters show more activity in the frontal and parietal cortices. |

| | | | | | |
|---|---|---|---|---|---|
| (Atherton et al., 2003) | 7 novices (M) | Solve chess positions | fMRI (14 ~ 16 sagittal slices) | Superior frontal lobes; parietal lobes; occipital lobes; left hemisphere | Bilateral activation is revealed in the superior frontal, parietal, and occipital lobes. Small areas of activation are observed unilaterally in the left hemisphere. The left hemisphere shows more activation than the right. |
| (Bilalić et al., 2010) | 8 experts (> 2000) 15 novices (M) | Recognition of chess positions | fMRI (176 sagittal slices) | Occipito-temporal junction | Experts' superiority in object recognition relates to bilateral activity next to the occipito-temporal junction. |

Table 5.1 *Cont.*

| Study | N | Tasks/paradigm | Method | Main brain areas/ structures | Main conclusion(s) |
|---|---|---|---|---|---|
| (Bilalić, Kiesel, et al., 2011) | 8 experts (~ 2130) 8 novices (M) | Geometric task; identity task; check task; eye movements | fMRI (176 sagittal slices) | Temporal gyrus; occipito-temporal junction; parieto-occipito-temporal junction; supra-marginal gyrus | Left temporal and parietal areas along the dorsal stream relate to chess-specific object recognition. Only in experts are homologous areas on the right hemisphere also engaged in chess-specific object recognition. |
| (Bilalić, Langner, et al., 2011) | 7 experts (~ 2100) 8 novices (M) | Recognition of faces and chess positions | fMRI (176 sagittal slices) | Fusiform face area (FFA) | Experts' FFAs are more activated than those of novices with naturalistic full-board chess positions and with randomly disturbed chess positions. |

| | | | | | |
|---|---|---|---|---|---|
| (Bilalić et al., 2012) | 8 experts (~ 2100)<br>15 novices<br>(M) | Recognition of chess position | fMRI (176 sagittal slices) | Dorsal stream; posterior temporal; left inferior parietal lobe; collateral sulci; bilateral retrosplenial | Experts' superior recognition performance and their functions arise in bilateral posterior temporal areas and the left inferior parietal lobe. The bilateral collateral sulci, together with the bilateral retrosplenial cortex, are more sensitive to normal than random positions among experts. |
| (Campitelli, Gobet, & Parker, 2005) | 2 experts (2550, 2450)<br>12 novices<br>(?) | Memorizing chess positions | fMRI (22 coronal slices) | Frontal lobes; posterior cingulate; cerebellum | There is brain activation in the frontal areas of the novices but not in the experts, who, rather, use from anterior to posterior areas of the brain. |

Table 5.1 *Cont.*

| Study | N | Tasks/paradigm | Method | Main brain areas/ structures | Main conclusion(s) |
|---|---|---|---|---|---|
| (Campitelli et al., 2007) | 5 players (~ 1971) (?) | Memorizing chess positions; recognition of chess positions | fMRI (22 coronal slices) | Temporal lobes; frontal and parietal lobes | The working memory tasks activate frontal and parietal lobes. Long-term memory tasks activate temporal areas. |
| (Campitelli et al., 2008) | 2 experts (2550, 2500) (?) | Memorizing chess positions | fMRI (22 coronal slices) | Left temporo-parietal junction; left frontal areas | The study finds a similar left-lateralized pattern of brain activity in both masters. The brain areas activated are the left temporo-parietal junction and left frontal areas. |

| | | | | | |
|---|---|---|---|---|---|
| (Duan et al., 2012) | 15 experts (2200 ~ 2600)<br>15 novices<br>(F = 6 experts, 6 novices) | Raven's Standard Progressive Matrices | fMRI (176 sagittal slices) | Caudate nucleus | Long-term chess training relates to smaller caudate nuclei, enhancing better integration of cognitive skill acquisition, in accordance with the default model network. |
| (Fuentes et al., 2018) | 1 expert (2550)<br>(M) | Chess game against a computer | EEG (theta Fz/alpha Pz) | Prefrontal cortex | Cortical theta Fz/alpha Pz ratio arousal increases and heart rate variability decreases during a chess game. The brain load increases during the chess game. There is pre-activation in a pre-game measure. The prefrontal cortex might be preparatorily activated. |

Table 5.1 *Cont.*

| Study | N | Tasks/paradigm | Method | Main brain areas/ structures | Main conclusion(s) |
|---|---|---|---|---|---|
| (Grabner et al., 2006) | • 47 experts (1325 ~ 2338)<br>• 23 high (~ 2076)<br>• 24 low (~ 1717)<br>• (M) | Personality (NEO-FFI); state anxiety (STAI); Intelligenz-Struktur-Test 2000; speed task (ST); memorizing chess positions; solving chess positions | EEG (event-related desynchronization; upper alpha band) | Parietal cortices; frontal cortices | Intelligence and chess expertise have different impacts on neural efficiency. More skilled chess players display higher activation over the parietal cortices and lower activation over the frontal cortices in the speed and reasoning tasks. |
| (Hanggi et al., 2014) | 20 experts (~ 2366)<br>20 non-players<br>(M) | Fluid intelligence; visuospatial abilities | MRI (160 sagittal slices) | Caudate nucleus; precuneus; occipito-temporal junction (OTJ) | Grey matter volume and cortical thickness are reduced in chess players compared with controls in the OTJ and precuneus. There are no differences in the volume of the caudate nucleus. |

| | | | | | |
|---|---|---|---|---|---|
| (Krawczyk et al., 2011) | 6 experts (~ 2515)<br>6 novices<br>(M) | Memorizing chess positions, faces, scenes, objects | fMRI (tilted axial slices: 3 mm thick, 0.5 mm slice gap) | Fusiform face area; cingulate cortex | Chess configurations are not strongly processed by face-selective regions. Areas in the posterior cingulate and right temporal cortex are more active in experts. The posterior cingulate cortex is responsive to chess only in experts. |
| (Li et al., 2015) | 29 experts (~ 2401)<br>29 novices<br>(F = 9 experts, 15 novices) | Age, sex, education, weight, handedness, mental and physical illness;<br>Raven's Standard Progressive Matrices | MRI (176 sagittal slices) | – | A multimodal MRI dataset. |

Table 5.1 *Cont.*

| Study | N | Tasks/paradigm | Method | Main brain areas/ structures | Main conclusion(s) |
|---|---|---|---|---|---|
| (Nichelli et al., 1994) | 10 (M) | Recognition of chess positions; solving chess positions | PET | Parieto-occipital lobe junction; left middle temporal gyrus; left superior premotor cortex; superior parietal lobe; medial superior parietal cortex; hippocampus; occipital-parietal junction; left orbitofrontal cortex; right prefrontal cortex | Solving a complex problem requires the activity of a network of several interrelated, but functionally distinct, cerebral areas. |

| | | | | | |
|---|---|---|---|---|---|
| (Onofrj et al., 1995) | 5 (1800 – 2200) (M) | Single chess position solving | SPECT | Frontal lobe | The non-dominant frontal lobe is active in the brain of chess experts when elaborating a solution for a complex chess problem. The chunking of elements into meaningful groupings and parsing of visual stimuli are functions of the non-dominant hemisphere. |
| (Powell, et al., 2017) | 12 novices (M) | Theory of mind (ToM) task; empathizing task; solving chess positions | fMRI (176 sagittal slices) | Temporo-parietal junction; superior temporal gyrus; posterior cingulate gyrus | ToM and empathy tasks activate the right-hemisphere orbitofrontal cortex and bilateral middle temporal gyrus. Chess tasks activate the medial-frontal and parietal cortex. Both areas overlap to a certain degree. |

Table 5.1 *Cont.*

| Study | N | Tasks/paradigm | Method | Main brain areas/ structures | Main conclusion(s) |
| --- | --- | --- | --- | --- | --- |
| (Rennig et al., 2013) | Data reanalysed from past studies (Bilalić et al., 2010, 2011, 2012) (M) | Recognition of chess positions | fMRI (176 sagittal slices) | Temporo-parietal junction | There is higher activation of the temporo-parietal junction in experts compared with novice players when presented with complex chess positions, suggesting that experts have higher visual integration skills. |

| | | | | | |
|---|---|---|---|---|---|
| (Saariluoma et al., 2004) | 6 (~ 2084) (M) | Blindfold chess tasks (memory and problem solving) | PET | Temporal lobe; frontal lobe | The memory task activates temporal areas; the problem-solving task activates frontal areas. Expert players process chess images differently from ordinary images. Visuospatial representations are characterized by large learned chunks and automated processing habits. |

Table 5.1 *Cont.*

| Study | N | Tasks/paradigm | Method | Main brain areas/ structures | Main conclusion(s) |
|---|---|---|---|---|---|
| (Silva-Junior et al., 2018) | 8 high-proficiency 8 low-proficiency (?) | Solving chess positions | EEG (entropy analyses, factor analyses) | Prefrontal cortex; occipital lobe; parietal lobe; temporal lobe | High-proficiency players present brain activation in areas related to planning and decision-making, such as the prefrontal cortex. Low-proficiency players show more brain activity in visual-processing areas, such as the occipital and parietal, or in areas related to eye control, such as temporal lobe association areas. |

| | | | | | |
|---|---|---|---|---|---|
| (Stepien, Klonowski, & Suvorov, 2015) | 3 experts (M) | Chess game against a computer | EEG (Higuchi fractal dimension; empirical mode decomposition) | – | The Higuchi fractal dimension is a better method for the analysis of EEG signals related to chess tasks than that of sliding window empirical mode decomposition. |
| (Volke et al., 2002) | 11 high-proficiency 11 low-proficiency (M) | Recognition of chess positions<br>Solving chess positions | EEG (coherence) | Frontal lobe; prefrontal cortex | Experts show well-mastered and automatic performance (high coherence), whereas novices show poorly mastered task resolution (low coherence). |

Table 5.1 *Cont.*

| Study | N | Tasks/paradigm | Method | Main brain areas/ structures | Main conclusion(s) |
|---|---|---|---|---|---|
| (Wright et al., 2013) | 14 experts (1650 – 2450) 14 novices (M) | Recognition of chess positions | EEG (N200, P300) | Frontal lobe | Expert–novice differences in posterior N200 begins early on check-related searches (240 ms). Prolonged N200 components reflect the matching of current perceptual input to memory, highlighting experts' superior pattern recognition and memory retrieval of chunks. |

et al., 2017). Regarding the brain-imaging method, fMRI and EEG were the most usual applications of choice, with eleven and six studies (44% and 24%), respectively. Two studies used MEG, PET, or MRI (8%), and only one study applied the SPECT method (4%).

The activity in most cerebral cortex areas and other brain structures was meaningfully related to the performance in the experimental tasks conducted in this set of studies. The frontal lobe was the brain area yielding the most interesting and remarkable findings independently of the applied method, however. The presentation of the main findings from this body of research is structured into the following four sections: cerebral cortex areas, hemispheric specialization, other brain areas and anatomical changes, and summarizing findings about brain functioning and chess.

## 5.5 Cerebral Cortex Areas

Prominent differences in the activation of the cerebral cortex areas between expert and novice chess players and in other domains emerge systematically (Bilalić, 2017). These differences have been mainly associated with working memory and long-term memory. For example, a review of PET and fMRI studies about expert memory when executing working-memory tasks highlights that the acquisition of expertise might run through a two-stage brain functional reorganization process (Guida et al., 2012). This reorganization process is here understood as the reallocation of brain resources occurring during expert acquisition from working memory to long-term memory. The studies with experts demonstrate brain activation of areas involved in long-term memory tasks, which suggests compatibility with functional brain reorganization. In contrast, the studies with novices demonstrate that there was brain deactivation of areas involved in working memory tasks, which suggests incompatibility with functional reorganization. This dual process is attributed to more practice through the acquisition of expertise. In this view, it is argued that the reorganization process contributes to increasing the size of the chunks, which would be stored in long-term memory, eventually leading to the consolidation of higher-level knowledge structures.

Additional fMRI studies with chess tasks requiring the use of memory have also underlined meaningful differences between the brain functioning of experts and novice chess players. For example, a study comparing two expert players with twelve novice players in the memorization of typical chess positions reports that novice players tend to use frontal lobe areas, whereas, in contrast, expert players use from anterior to posterior areas of the brain (Campitelli et al., 2005). Another study with five expert players informs that frontal and parietal brain areas are activated in working-memory tasks, whereas temporal areas are activated in long-term memory tasks (Campitelli et al., 2007). These findings support the view that chess players employ long-term memory chunks, which are purportedly stored in the temporal lobes (Gobet & Simon, 1996c).

Furthermore, the advantage of experts over novice players is highly consistent in the recognition of chess positions regarding the brain activity next to the occipito-temporal junction (Bilalić et al., 2010), and posterior temporal and left inferior parietal lobes. Moreover, the collateral sulci and retrosplenial cortex have also been shown to be more sensitive among experts than among novices (Bilalić et al., 2012). These latter findings about the recognition of chess objects and their interrelationships evidence the extensive knowledge of expert players about specific chess patterns. The performance in recognizing chess patterns drops meaningfully with the randomization of meaningless chess positions, however. Furthermore, experts use the same regions as novices do when recognizing chess objects, though experts may also use additional brain regions. There appears to be, therefore, a learning process, whereby experts would be at advanced stages of learning, whereas novices might be at earlier stages of learning. Further reanalyses of some of these fMRI studies have focused on the temporo-parietal junction, a multimodal sensory brain area with a principal role in the social brain (Rennig et al., 2013). This brain area is more activated in expert than in novice players when exposed to complex chess positions, suggesting higher visual integration skill on the part of experts vis-à-vis novices.

Other differential activation patterns between expert and novice chess players, mainly located in the frontal and temporal cerebral areas, have also been found employing other brain-imaging techniques, such as EEG, fMRI, or PET. For example, EEG findings indicate that better performers show brain activation in the prefrontal cortex, whereas worse performers show brain activation in visual-processing brain areas, such as the temporal, occipital, and parietal areas, during a chess problem-solving task (Silva-Junior et al., 2018). Similarly, there is support for the neural efficiency hypothesis when analysing the brain activity of a considerable group of chess players (Grabner et al., 2006). Higher-skilled players show a lower level of activation in the frontal cortices but a higher level of activation in the parietal cortices when compared with lower-skilled players in short-term memory and reasoning tasks.

A particularly attractive technique is that involving focal gamma bursts obtained with the MEG technique. The oscillatory neural activity in the gamma frequency band (30 to 100 Hz) appears at early stages less than 150 ms after stimuli onset and at later stages more than 200 ms after stimuli onset. This signal is particularly noteworthy for the cognitive processes involved in chess playing, because it implies several cognitive properties that are useful in chess. It has been consistently linked with attention and perceptual processes, motor tasks, short- and long-term memory, and problem solving (Rieder et al., 2011). Furthermore, it has also been associated with more specific cognitive functions such as object representation, memory and language processes, visual awareness, and for the performance in memory tasks implying subsequent recognition. In addition, the match-and-utilization model (MUM) attempts to explain the gamma-band oscillation in terms of two main processes: comparing memory material with the stimuli, and the utilization of signals obtained from this comparison (Herrmann, Fründ, & Lenz, 2010).

The gamma-band activity predominates in different cortical areas when contrasting expert with novice chess players. Two studies report that focal gamma bursts were more frequently observed in the frontal and parietal lobes of chess grandmasters, whereas, in contrast, focal gamma bursts were more frequently observed in the deeper structures of the medial temporal lobes of novice chess players (Amidzic et al., 2006, 2001). These findings are interpreted in accordance with memory formation in novice chess players. The medial temporal lobe might play a transitional role during the creation of expert memory. In this view, expert chess players recall stored information from long-term memory, whereas novice chess players need to encode the new information. This discrepancy is argued as supportive of the theory regarding the extensive repository of chunks stored in experts' long-term memory – i.e., templates.

In blindfold chess, looking at the pieces on the board is not allowed, and the opponent's moves are transmitted verbally; hence, blindfold chess is particularly appealing for examining taxing brain functions involving memory and problem solving. The demanding task of blindfold chess was examined with the PET method (Saariluoma et al., 2004). Different tasks elicited the activation of different cerebral cortex areas. A memory task activated temporal areas, whereas a problem-solving task activated frontal areas. These findings uphold the notion that the visual representations of expert players were guided by learned chunks of chess information, and by highly automated information-processing habits. Bearing in mind the severe restriction imposed by the blindfold chess paradigm, these outcomes support the view that chess-related visual information is probably represented in the brain in a quite different way from regular images because of the previously learned visuospatial chunks, which might be automatized to a great extent and retrieved from long-term memory.

## 5.6 Hemispheric Specialization

Crucial qualities of the human brain are cerebral lateralization and hemispheric specialization. The left hemisphere dominates linguistic functioning in most people, whereas the right hemisphere is more important in the experience and expression of emotions (Clark et al., 2010). Some studies examining the brain functioning of chess players have addressed different aspects of the hemispheric specialization issue. For example, a divided-visual-field experiment about accuracy performance and reaction time highlights brain hemispheric specialization regarding the perceptual organization of chess information (Chabris & Hamilton, 1992). Within this paradigm, there should be a right hemisphere advantage for single-chunk fragments of chess information, whereas there should be a left hemisphere advantage for multiple-chunk fragments of chess information. The findings from sixteen right-handed master-level chess players indeed suggest that the right hemisphere is more efficient at identifying chess positions with a single chunk configuration, while the left hemisphere is more efficient at identifying chess positions with multiple-chunk configurations. Because of the right hemisphere specialization concerning complex visuospatial tasks, these findings support the notion of the superiority of the right hemisphere at learning chess chunking rules.

These latter findings also parallel the outcomes from an fMRI study with just two right-handed grandmaster players, whereby the left temporo-parietal and left frontal areas were activated when confronted with an autobiographical memory task (Campitelli et al., 2008). That both players displayed such analogous left brain hemisphere lateralization corroborates the superiority of expert over novice players at recovering chess chunks and in pattern recognition, which was additionally substantiated by EEG findings (Wright et al., 2013). Furthermore, the larger amount of knowledge gained by experts over novices has been argued to be the main underlying factor of the principle of the 'double take' of expertise (Bilalić, 2017, 2018). In accordance with this principle, the crucial neural difference between experts and novices is that, although both groups activate similar brain areas in executing a cognitive task, experts tend to activate additional homologous areas in the opposite brain hemisphere.

Higher left hemisphere activation has also been found with novice players when solving chess positions (Atherton et al., 2003). This fMRI finding was detected mainly in the occipital and parietal lobes, however. In contrast, the frontal lobes were unusually deactivated, suggesting that certain problem-solving skills implicated in chess would depend on cerebral cortex areas other than the lateral prefrontal cortex, and particularly anticipated at lower levels of chess skill. Furthermore, the SPECT study carried out with five expert players highlights the importance of the dorsal prefrontal and middle temporal lobes in the non-dominant hemisphere for chess skill: the right hemisphere for four right-handed players, and the left hemisphere for the one left-handed player (Onofrj et al., 1995). The hemispheric non-dominant frontal lobe, and to a lesser extent the temporal lobe, were activated when attempting to find a solution for a complex chess problem, suggesting that the chunking of chess elements into meaningful patterns and parsing of visual stimuli are functions governed by areas located in the non-dominant brain hemisphere. Additional fMRI findings suggest that left hemispheric temporal and parietal areas activated when recognizing chess-specific objects are also activated in the right hemisphere only for experts (Bilalić, Kiesel, et al., 2011).

## 5.7 Other Brain Areas and Anatomical Changes

Sophisticated brain-imaging techniques allow us to examine other brain structures and functions. One such structure is the fusiform face area (FFA), a visual area located in the temporal lobe that is activated even at extremely young ages when viewing faces (Clark et al., 2010). The FFA is also activated when exposed to objects entailing a certain experience for the individual, such as birds for an ornithologist (Bukach, Gauthier, & Tarr, 2006).

The FFA was analysed with seven chess experts and eight chess novices through the fMRI technique by contrasting the recognition of faces and chess positions (Bilalić, Langner, et al., 2011). In the recognition of both kinds of objects, the FFA of experts was more activated than that of novices when viewing both coherent chess positions and randomly disrupted positions. These findings support the view that the specific nature of the objects within a given expert domain might modulate the FFA's activity. In addition, contrasting the recognition of faces, scenes, objects, and chessboards was investigated in another fMRI study by comparing six experts with six novice chess players. This study involved the analyses of brain regions involved in the processing of faces and chess configurations, to ascertain whether the perception of both kinds of objects activates common brain regions (Krawczyk et al., 2011). The main findings in this study suggest that face and chess processing occurs independently. Face-processing areas such as the FFA are disengaged from the processing of chess stimuli. For example, a notable activated brain area in the expert players was the posterior cingulate cortex, corroborating the memory retrieval and thinking requirements during the performance of chess tasks, as reported elsewhere (Campitelli et al., 2007).

Brain areas related to the theory of mind (ToM), such as the temporoparietal junction, superior temporal gyrus, and posterior cingulate gyrus, have also been examined in relation to chess playing (Powell et al., 2017). ToM is the ability to understand and predict the mental state of other people, including emotions, desires, and intentions. This attribute may be quite important in chess playing when attempting to anticipate the tactical and strategic plans of the opponent. Twelve novice players were exposed to ToM, empathy, and chess-problem-solving tasks while their brain activity was recorded with fMRI. Brain areas involved in ToM were mainly activated with ToM and empathy tasks, whereas cortical areas in the medial frontal and parietal lobes were activated in turn with chess tasks. A certain degree of neural overlap between both processes is also reported in this study, however. Making decisions about potential chess moves intertwined with ToM considerations such as reasoning iteratively about the opponent's moves.

Another intriguing feature concerning the inter-individual variability between chess player's brains is concerned with whether there are brain anatomical changes associated with the level of chess skill. Anatomical changes have been reported to occur in two complex brain structures, the precuneus and the caudate nucleus. The precuneus is located in the superior parietal lobe. This structure exhibits alternate connections with the frontal lobe and is involved with functions such as consciousness, body movements, self-awareness, episodic memory retrieval, and visuospatial imagery. The caudate nucleus is located in the basal ganglia. This is a group of grey matter nuclei at the base of the brain hemispheres closely related to the frontal lobes in the acquisition and expression of cognition. The caudate nucleus plays a crucial role, for instance, in the serial order of movements and behaviour (Clark et al., 2010). Both the precuneus and the caudate nucleus have been linked with the search and generation of moves in professional players of *shogi*, a Japanese game sharing many characteristics with conventional chess (Wan et al., 2011).

Regarding the anatomical changes in both structures in chess players, the caudate nuclei are reported as being meaningfully smaller in expert than in novice players (Duan et al., 2012). Expert players displayed a larger default brain network than novices, however. A default brain network is a concept that encapsulates a group of interconnected brain structures that are active in a baseline resting state, namely the caudate nucleus, the posterior cingulate cortex, and the angular gyrus. These two apparently contradictory findings are interpreted in terms of a synaptic pruning mechanism. The removal of superfluous synapses would contribute to an enhanced and more efficient integration of brain functioning across different brain areas and structures.

These findings are partially corroborated in a comprehensive neuroanatomical MRI study with twenty expert chess players compared with twenty non-player controls (Hanggi et al., 2014). This study applied diffusion tensor imaging (DTI) and voxel-based morphometry (VBM) and surface-based morphometry (SBM) techniques to determine the grey matter volumetric characteristics. Grey matter volume in the precuneus and cortical thickness in the occipito-parietal junction were lower in chess players than in control subjects. On the other hand, grey matter volume in the caudate nucleus was very similar in chess players and controls. In accordance with a synaptic pruning mechanism, this outcome is attributed to the earlier starting age for chess playing by the experts involved in the study: between four and fourteen years of age (M = eight years old). Because sensory deprivation during childhood appears to reduce synaptic pruning to a greater extent, stimulation in the chess domain at such young ages might have stimulated a remarkable removal of synapses. This interesting study raises several supplementary questions related to the observed inter-individual variability, however. For example, were the differences in anatomical characteristics the cause or the consequence of becoming involved in chess to the extent of reaching expert level?

## 5.8 Summarizing Findings about Brain Functioning and Chess

The evidence found in this body of research points to the meaningful activation of several areas in the cerebral cortex during a variety of experimental tasks. Brain areas show a notable degree of specialization concerning specific tasks when dealing with chess information. In addition, there are remarkable inter-group differences in brain activity when comparing experts with novice players. For example, diverse anatomical structures and specialized brain areas appear to function as an integrated synchronized network that might be enhanced at higher levels of expertise and chess skill (Nichelli et al., 1994; Volke et al., 2002). Moreover, anticorrelated functional networks may contribute to accounting for these findings. When performing attention and cognitive tasks there are brain areas with increased activity and there are brain areas with decreased activity. Typically, frontal and parietal cerebral lobes areas show increased activity, while the posterior cingulate, medial and lateral parietal, and medial prefrontal cortices show decreased activity. This imbalance grows with progressively more demanding task requirements (Fox et al., 2005).

This chapter has addressed the biological underpinnings of chess playing and chess expertise. Psychophysiology and brain activity are fascinating approaches to addressing individual differences using chess as a model domain. Playing chess implies a demanding learning process, and expert chess players reach surprisingly higher performance levels than non-expert players do. This advantage is also reflected in the brain functioning and brain structure of chess players, because there are neuroanatomical and neurophysiological skill adaptations on the part of experts. The body of evidence reviewed in this chapter reflects well six main themes that emerge when explaining how learning a given skill might impinge on substantial observable brain changes (Hill & Schneider, 2006). These six themes are the following: (1) learning is localized and specialized; (2) learning and processing occur in similar brain locations; (3) learning improves processing; (4) some tasks can be reorganized; (5) domain-meaningful stimuli are processed in a special way by experts; and (6) learning produces observable changes. Concerning the specific body of research into the psychophysiology and brain functioning of chess players, there are a few main points that can be singled out for highlighting.

1. Chess experts display different brain activation patterns from novice players.
2. Chess experts appear to be able to integrate more brain areas when dealing with chess-related tasks, such as recognition, memorizing, and problem solving.
3. Expert chess performance implies higher brain hemispheric specialization. Chess experts are more prone to use the non-dominant brain hemisphere than novice players are.

4. Brain activation and brain structural differences between chess players of different levels of skill could be due to a developmental process because of starting to play chess very early in life, and devoting highly varying periods to studying the extensive knowledge base within the chess domain.
5. There are two main limitations on this body of knowledge. First, the low sample size impairs the statistical power of the studies. Second, most of the samples used in these kinds of studies are males, with an overall male to female ratio (M:F) of 11:1. This disparity may bear a remarkable sex bias, precluding the generalization of these findings to the female population.
6. Moreover, there is a paucity of studies regarding the psychophysiological parameters and hormonal activity of chess players. This relatively unexplored field could be examined in greater depth by incorporating measures of individual differences in a variety of psychological traits related to chess performance.

In the light of the current findings concerning the brain functioning of chess players, one question that may be raised is whether chess players are more intelligent than average people are. With better integration of activity in several brain areas, it might be the case that the answer to this question is affirmative. The study of intelligence in chess has yielded inconclusive and sometimes controversial findings, however.

6

# Intelligence

Intelligence is probably the most important construct in psychology. Ever since the seminal work at the beginning of the twentieth century suggesting that intelligence is a very general ability, much has been said about human intelligence (Spearman, 1904, 1927). There are several formal models providing a comprehensive account of intelligence, a psychological abstract concept that has important scientific and political implications (Sternberg & Kaufman, 2011). The scores in intelligence tests relate strongly to several variables with a social and practical relevance for daily life. There are robust positive correlations between the intellectual level, measured with psychometric intelligence tests, with work performance, academic achievement, and economic success. Conversely, these correlations tend to be negative with unemployment, delinquency, disease, and mortality. No other psychological variable produces such correlations.

The publication in 1994 of the book *The Bell Curve* elicited a cogent debate in the mass media about human intelligence (Herrnstein & Murray, 1994). This debate was at times driven by significant misinterpretations regarding the extant scientific body of knowledge about human intelligence. Consequently, many scholars and researchers in the field reacted in an attempt to clarify several of the most crucial aspects about human intelligence. For example, the Board of Scientific Affairs (BSA) of the American Psychological Association (APA) encouraged the elaboration of a report about the meaning of test scores and the nature of intelligence (Neisser et al., 1996). This report raised several matters that were unresolved at that time, including the following:

1. the mechanisms by which individual differences in genes contribute to individual differences as measured by psychometric tests, and, to a greater extent, at older ages;
2. the influence of environmental factors in the development of intelligence, particularly of schooling;
3. the role of nutrition;
4. the pattern of the relationships between measures of information-processing speed and psychometric intelligence;
5. the progressive and generalized increment of means in intelligence tests in the past fifty years;

6. the differences in mean intelligence test scores between racial groups (i.e., blacks and whites), which are unsupported by either genetic or environmental factors; and
7. the measurement and nature of other constructs akin to intelligence, such as creativity and wisdom.

Another similar report around that time highlighted mainstream conclusions concerning the nature, origins, and consequences of individual and group differences in intelligence (Gottfredson, 1997b). Fifty-two experts in intelligence and related fields signed this report, which is structured around six main themes: meaning and measurement; group differences; practical importance; the sources and stability of within-group differences; the sources and stability of between-group differences; and the implications for social policy. Further work has summarized the main opinions when describing the construct of intelligence (Hunt & Carlson, 2007):

1. intelligence does not exist;
2. intelligence does exist and is measured by intelligence tests;
3. intelligence does exist and is not measured by intelligence tests;
4. intelligence is unchangeable, fixed at birth;
5. intelligence is in part determined by the environment, especially through education;
6. intelligence overlaps with learning ability;
7. intelligence is purely cognitive; and
8. intelligence can take many forms, in domains as diverse as music, mathematics, athletics, and leadership.

As with the work by Earl Hunt, the approach taken in this book considers that intelligence can be viewed as the combination of points 2, 5, 6, and 7. Intelligence exists and it can be measured with intelligence tests; it is determined by both genetic and environmental factors; it is intimately related to the ability to learn; and it can be constrained to the cognitive domain.

Because the conceptualization and measurement of intelligence has relied on specific tests, several psychometric instruments have been designed to gauge a variety of cognitive abilities. Two desirable properties of intelligence tests are reliability and validity. A test is reliable when it provides the same measurement on different occasions. A test is valid when it measures the construct that it is intended to measure and is predictive of other constructs. In general, tests designed to evaluate cognitive abilities meet both these properties to a large extent. Cognitive abilities tests can be classified in four main ways (Urbina, 2011): (1) examination mode (individual or group); (2) population (children, adults, specific groups); (3) content (verbal, non-verbal); and (4) length (full, abbreviated). Table 6.1 shows an overview of some of the most popular tests to evaluate human cognitive abilities in accordance with the

Table 6.1 *Overview of some tests to evaluate cognitive abilities*

| Test | Reference | Examination | Broad factors/subtests |
|---|---|---|---|
| Armed Services Vocational Aptitude Battery (ASVAB) | – | Group | General science, arithmetic reasoning, word knowledge, paragraph comprehension, mathematics knowledge, electronics information, auto and shop information, mechanical comprehension, object assembly |
| Cognitive Abilities Test (CAT-3) | (Lohman & Hagen, 2002) | Group | Verbal, quantitative, non-verbal |
| Cognitive Assessment System (CAS) | (Naglieri & Das, 1997) | Individual | Planning, attention, simultaneous processing, sequential processing |
| Differential Aptitude Test (DAT-5) | (Bennett, Seashore, & Wesman, 2002) | Group | Verbal reasoning, numerical ability, abstract reasoning, mechanical reasoning, space relations, language use |
| Fagan Test of Infant Intelligence (FTII) | (Fagan & Detterman, 1992) | Individual | Visual recognition memory |
| Kaufman Adolescent and Adult Intelligence Test (KAIT) | (Kaufman & Kaufman, 1993) | Individual | General IQ, crystallized intelligence (*Gc*), fluid intelligence (*Gf*) |
| Otis-Lennon Test | – | Group | Verbal comprehension, verbal reasoning, figural reasoning, quantitative reasoning |
| Peabody Picture Vocabulary Test (PPVT-4) | (Dunn & Dunn, 2007) | Individual | Vocabulary ability |
| Raven's Progressive Matrices | (Raven & Raven, 2008) | Group | General intelligence (*g*) |

Table 6.1 *Cont.*

| Test | Reference | Examination | Broad factors/subtests |
|---|---|---|---|
| SAT | sat.college-board.org | Group | Writing, reading, mathematics |
| Stanford-Binet 5 | (Roid, 2003) | Individual | Fluid reasoning, knowledge, quantitative reasoning, visual-spatial processing, working memory |
| Wechsler Adult Intelligence Scale-Fourth Edition (WAIS-IV) | (Wechsler, 2008) | Individual | General IQ, verbal comprehension, perceptual reasoning, working memory, processing speed |
| Wechsler Intelligence Scale for Children-IV (WISC-IV) | (Wechsler, 2003) | Individual | General IQ, verbal comprehension, perceptual reasoning, working memory, processing speed |
| Wonderlic Personnel Test (WPT) | (Wonderlic & Wonderlic, 1992) | Group | Verbal, arithmetic, and logical questions |
| Woodcock-Johnson III Test of Cognitive Abilities (WJ III) | (Woodcock, McGrew, & Mather, 2001) | Individual | Achievement tests (22), cognitive tests (20) |

*Note:* The 'SAT' acronym initially stood for Scholastic Aptitude Test, though it has been renamed several times. This test is used extensively as part of the college admission process in the United States.

examination mode, and lists the cognitive abilities or performances evaluated by each instrument. Some tests comprise several subtests addressing more specific cognitive abilities or performances, such as the ASVAB or the WAIS-IV. Other tests measure just one broad ability or performance, such as the PPVT-4 or Raven's Progressive Matrices.

A popular score of intelligence conceived in the early twentieth century is the intelligence quotient (IQ), expressed as mental age divided by chronological age (Stern, 1921). The IQ is generally used to provide a quantification of the level of intelligence, which can be readily obtained from several of the tests used to evaluate cognitive abilities. Because intelligence tests are standardized with

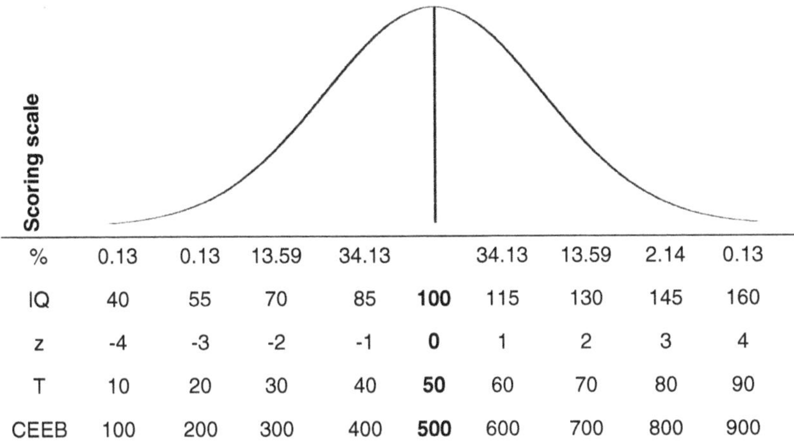


**Figure 6.1** The normal distribution with IQ scores compared with the approximate percentage of cases under the curve, and other scoring systems
*Notes*: % = approximate percentage of cases from the population under different areas of the curve; IQ = intelligence quotient scores; z = standardized scores; T = T scores; CEEB = College Entrance Examination Board.

a representative sample of people, the IQ is usually gauged with an arbitrary mean of 100 IQ points and a standard deviation of 15 IQ points. The IQ follows a normal distribution fairly well, as shown in Figure 6.1. Over 68% of the population falls between one standard deviation below and above the mean, whereas the number of people at both ends of the IQ continuum decreases progressively. Identifying intelligence with the IQ may not be appropriate, however. The IQ is an individual score obtained in a test relative to a broader group of people. In contrast, intelligence is a complex multi-factorial concept with an intricate network of causes and consequences conceived as part of an extensive open system. In this view, intelligence is unobservable; it is, rather, a concept built in accordance with how it is envisaged (Hunt, 2011). For example, the Flynn effect is a remarkable observed phenomenon maintaining that IQ scores in intelligence tests rose over the twentieth century across different cultures (Flynn, 1984, 1987). A striking finding about the Flynn effect, however, suggests that IQ scores measure in fact a correlate that relates weakly to intelligence.

## 6.1 Approaches to the Study of Intelligence

There are several ways to look at what human intelligence means. Historically, diverse theories have been proposed for conceptualizing and accounting for what intelligence refers to. Theories and models about intelligence can be classified into three broad levels of analysis: psychometric, information processing, and biological (Hunt, 2011), even though other authors may employ other classification schemes, such as psychometric, physiological, and social (Davidson & Kemp, 2011). Whatever the case, when it comes to investigating the structure, causes, and consequences of human intelligence some empirical studies may approach the study of intelligence from more than one level of analysis. An example could be to examine whether a general intelligence model (psychometric level) predicts the premises posited by the P-FIT model (biological level) concerning brain structure (Colom et al., 2009). Another example could be to analyse whether a measure of general intelligence obtained with Raven's Progressive Matrices (psychometric level) relates to a reaction time (information-processing level) measure (Jensen & Munro, 1979). Table 6.2 shows the most meaningful theories and models that

Table 6.2 *Theories and approaches to the study of human intelligence*

| Level | Theory/model/paradigm |
|---|---|
| Psychometric | Spearman's *g* |
| | Vernon's *v:ed* and *k:m* |
| | Thurstone's primary mental abilities |
| | Guilford's structure of intellect |
| | Fluid (*Gf*) and crystallized (*Gc*) intelligence |
| | Hierarchical LISREL (HILI) |
| | Carroll's three-stratum |
| | Cattell–Horn–Carroll (CHC) |
| | *g*-VPR |
| Psychometric extensions/social | Planning, attention, simultaneous, successive (PASS) |
| | Triarchic theory |
| | Multiple intelligences |
| | PPIK |
| | Emotional intelligence |
| Information processing | Speed of mental processing (RT, IT) |
| | Working memory |
| | Verbal comprehension |
| | Visual-spatial reasoning |
| Biological | Neural efficiency hypothesis (NEH) |
| | P-FIT |
| | Neural plasticity |
| | EDSC |

*Notes: v:ed* = verbal, educational; *k:m* = spatial, practical, mechanical; *g*-VPR = verbal, perceptual, rotation (manipulation of visual objects); PPIK = intelligence as process, personality, interests, intelligence as knowledge; RT = reaction time; IT = inspection time; P-FIT = parieto-frontal integration theory; EDSC = ecological dominance–social competition.

have been delineated since the beginning of the scientific study of human intelligence. These are classified at four differentiated levels of analysis: psychometric; psychometric extensions or social; information processing; and biological.

Theories at the psychometric level are mainly concerned with the structure of human cognitive abilities, relying on a statistical technique known as factor analysis. Factor analysis aims to explain the covariances among several observable measures in terms of a significantly lower amount of latent or unobservable dimensions. When submitting a number of test scores of diverse cognitive abilities to a factor analysis, a recurrent observation is that all measures show meaningful positive correlations. This outcome is known as the positive manifold, and it is commonly admitted as consistent evidence of an underlying general factor of intelligence. The positive manifold dates back to the earliest work of human intelligence with the factor analysis technique (Spearman, 1904; van der Maas et al., 2006). Some psychometric models represent a hierarchy with the general factor of intelligence (*g*) at the highest level, broad cognitive abilities underneath *g*, and narrower abilities at the next level, such as John Carroll's three-stratum (Carroll, 1993). Other psychometric models suggest that human intelligence is non-hierarchical, with several cognitive abilities placed at the same explanatory level, such as Louis Thurstone's primary mental abilities model (Thurstone, 1938). Figure 6.2 compares the factor structures of both models.

The fluid and crystallized intelligence model (*Gf* ~ *Gc*) is another influential psychometric theory of human intelligence (Cattell, 1963, 1987). Fluid intelligence is tapped by inductive, deductive, and quantitative reasoning, representing the ability to tackle novel problems quickly and efficiently, and is thought to underlie the biological basis of human intelligence. Crystallized intelligence is tapped by tests dealing with general knowledge, the use of language, and a variety of learned skills, and is thought to underlie the environmental basis of human intelligence. Extensions of psychometric models respond to some sort of disappointment with cognitive abilities tests (Hunt, 2011). These extensions attempt to conceptualize intelligence as related to central realms of human activity, such as mental health and functioning (Das, 1999), education (Gardner, 1993), or overall success in life (Sternberg, 1999). A remarkable approach is that offered by the intelligence as process, personality, interests, and intelligence as knowledge theory (PPIK: see Figure 4.5). This theory is noteworthy because it is a comprehensive, integrative account for explaining intellectual development into maturity by outlining the interplay of other individual traits, such as personality, interests, and domain knowledge (Ackerman, 1996; Ackerman & Heggestad, 1997).

The scientific study of intelligence at the information-processing level aims to examine the basic processes of intelligence that underlie individual differences in the central nervous system that are influential on the speediness in decision-making (Nettelbeck, 2011). Several elementary cognitive tasks

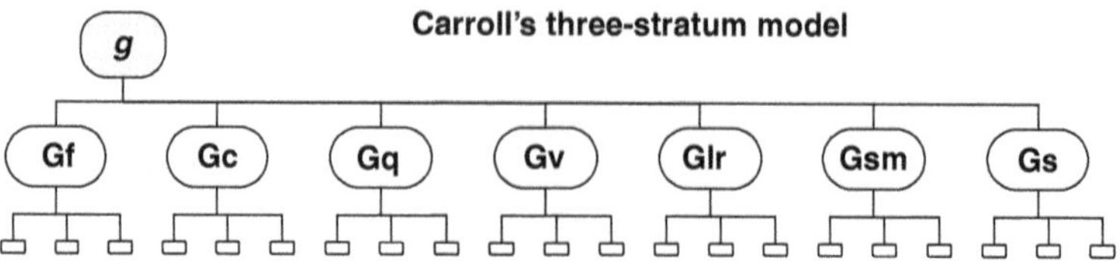


g = general intelligence; Gf = fluid intelligence; Gc = crystallized intelligence;
Gq = quantitative ability; Gv = visual-spatial ability;
Glr = long-term storage and retrieval in memory;
Gsm = short-term memory; Gs = cognitive speediness

**Thurstone's primary mental abilities model**

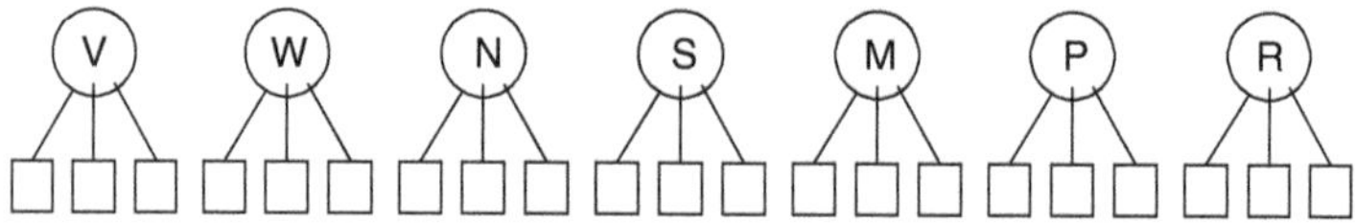


V = verbal comprehension; W = verbal fluency; N = numerical ability; S = spatial ability;
M = memory; P = perceptual speed; R = inductive reasoning

**Figure 6.2** Hierarchical (Carroll's model) and non-hierarchical (Thurstone's model) psychometric models of human intelligence; the squares in both models represent the specific tests used to measure each broad factor

(ECTs) are designed to tap mental processing speed in accordance with distinct paradigms, such as reaction time (RT) and inspection time (IT), working memory, verbal comprehension, and visual-spatial reasoning processes. The performance in an ECT under one or more of these paradigms correlates with cognitive abilities calibrated with psychometric tests. A consistent body of evidence points to a significant negative correlation between speed of mental processing (i.e., RT, IT) with psychometric measures of cognitive abilities. People with higher scores in psychometric tests tend to be faster at information processing (Deary & Stough, 1996; Grudnik & Kranzler, 2001; Jensen & Munro, 1979; Johnson & Deary, 2011; Nettelbeck & Lally, 1976).

Because ECTs appear to underpin individual differences in the central nervous system, information-processing theories of intelligence have been combined with psychometric and psychophysical approaches, bridging psychometric and biological theories of intelligence (Deary, 2001; Hunt, 2011). For example, the aforementioned neural efficiency hypothesis of intelligence maintains that people with higher scores in psychometric measures of cognitive abilities tend to show lower brain activation when performing a cognitive task than people with lower scores, who tend to show higher brain activation (Neubauer & Fink, 2009a). On the other hand, the parieto-frontal integration theory of intelligence accounts for the association between individual differences in psychometric cognitive abilities with variations in brain structure and functioning (Jung & Haier, 2007). The P-FIT contends that an integrated brain network comprising predominantly areas in the parietal and frontal lobes, but also the anterior cingulate gyrus and regions in the temporal and occipital lobes, organizes the brain foundation of human intelligence.

Intelligence is the psychological attribute with the greatest influence on central realms of human activity, such as work, health, and education, and also for everyday functioning in a variety of domains (Gottfredson, 1997a, 2004), but how about chess? Is a high level of intelligence a necessary requirement to perform well in the ultimate intellectual game? Are chess players more intelligent than the average population? Are stronger/expert players more intelligent than weaker/novice players are?

## 6.2 Individual Differences in Intelligence and Chess

Chess is the prototypical intellectual game, frequently associated with intelligence in the media and society. Undoubtedly, playing chess is a demanding intellectual activity. As in other fields involving the management of complex information and abstract relationships, chess imposes an intensive use of several cognitive abilities. Earlier studies about the psychology of chess concurred reasonably well about the essential mental qualities of chess master players. Apart from youth and physical robustness, the seminal work by Binet in the late nineteenth century highlighted the importance of memory or mental calculation (Binet, 1894). Alfred Cleveland also claimed that a persistent chess memory, quickness of perception, constructive imagination, accurate and deep analysis skills, and a general mental ability were all desirable attributes for performing well in chess, while highlighting a high level of chess performance as being compatible with success in other intellectual areas (Cleveland, 1907). Cleveland also acknowledged, however, that the aforementioned abilities were rather constrained to the chess domain, and that chess skill should not be taken as a valid indicator of mental endowment (305). Besides, logical thinking, calculation speed, and imagination and will were deemed to be crucial preconditions for reaching a high level of chess performance (Djakow et al., 1927). On the other hand, de Groot highlighted that chess skill is largely determined by spatial, verbal, and learning abilities, and by an extensive knowledge base acquired from individual experience (de Groot, 1965). In addition, de Groot delved into the relationship between chess talent and mathematical talent, and the personality types among chess masters, by emphasizing the 'extracurricular achievements of chess masters'. In Table 15 of his work (362–364), de Groot summarizes the chess status, training, and profession of 55 renowned grandmasters at that time. Some 24% of the players were involved in training and professions related to mathematics or hard science fields. This finding was justified because of the parallels between mathematical and chess thinking, such as spatial reasoning and mental flexibility. In contrast, however, it was also argued that, unlike mathematical thinking, chess thinking also requires permanent mental productivity, together with intuitive judgements, and decision-making under circumstances with incomplete information.

Increasingly young players have dominated the world of top-level chess since the early 1970s by outperforming older players. This fact has been taken as real-world evidence for the growth in intelligence, put forward as the Flynn effect (Flynn, 1984, 1987; Howard, 1999). There are alternative explanations for the growing dominance of younger players in elite chess, however (Gobet, Campitelli, & Waters, 2002). Younger players would outperform older players not through being more intelligent but because of having more available chess information and new coaching methods, playing more frequently, starting to play earlier, and the introduction of new chess regulations concerning thinking time and tournament schemes. Further analyses with a wealth of chess data have argued that, since 1970, chess players really have achieved higher chess performance levels at progressively younger ages, however (Howard, 2005b). This age effect, observed in this latter study, was more robust within top players, which might suggest natural talent as a source of strength in accounting for individual differences in chess skill over the aforementioned changes in the chess environment.

Table 6.3 summarizes similar biographical information to that provided by de Groot and Howard for unofficial and official world chess champions. It covers a span of 258 years, from François-André Danican Philidor in 1755 to Magnus Carlsen in 2013 (de Groot, 1965; Howard, 1999). This list includes leading players who were considered at the time the best chess players in the world, though the first undisputed world chess champion was Wilhelm Steinitz in 1886. For each of these twenty-six individuals, Table 6.3 displays the country of origin, the year of birth (B), the year he first won a world championship (W), the age of becoming world champion (W–B), and other intellectual activities. This information has been mainly collected from previous work from de Groot and Howard and from current specialized web pages (de Groot, 1965; Howard, 1999).

Even though the number of individuals is considerably lower than that reported by de Groot, these data underline two interrelated points worth mentioning. The first point is that there are fewer intellectual or professional activities other than chess in later years. There are more players with

Table 6.3 *Unofficial and official world chess champions and additional intellectual activities*

| Champion (*) | Country | Born (B) | Win (W) | Δ(W–B) | Intellectual activities |
|---|---|---|---|---|---|
| François-André Danican Philidor | France | 1726 | 1755 | 29 | Music composer |
| Alexander Deschapelles | France | 1780 | 1815 | 35 | Military career |
| Louis de La Bourdonnais | France | 1795 | 1834 | 39 | Chess writer |
| Howard Staunton (2) | UK | 1810 | 1843 | 33 | Chess journalist |
| Adolf Andersen (2) | Prussia | 1818 | 1851 | 33 | Mathematician, high school teacher |
| Paul Morphy | US | 1837 | 1858 | 21 | Degree in law |
| Wilhelm Steinitz (5) | Austria | 1836 | 1866 | 30 | Chess journalist |
| Emmanuel Lasker (6) | Prussia | 1868 | 1894 | 26 | Doctorate in mathematics |
| José Raúl Capablanca | Cuba | 1888 | 1921 | 33 | Chemical engineering student |
| Alexander Alekhine (4) | Russia | 1892 | 1927 | 35 | Degree in law, chess writer |
| Machgielis Euwe | Netherlands | 1901 | 1935 | 34 | Doctorate in mathematics |
| Mikhail Botvinnik (5) | Russia | 1911 | 1948 | 37 | Doctorate in electrical engineering |
| Vasily Smyslov | Russia | 1921 | 1957 | 36 | Opera singer, chess writer |
| Mikhail Tal | Latvia | 1936 | 1960 | 24 | Literature degree, chess writer |
| Tigran Petrosian (2) | Georgia | 1929 | 1963 | 34 | Chess journalist |
| Boris Spassky | Russia | 1937 | 1969 | 32 | Journalism degree |
| Robert Fischer | US | 1943 | 1972 | 29 | Patent author |

Table 6.3 *Cont.*

| Champion (*) | Country | Born (B) | Win (W) | Δ(W–B) | Intellectual activities |
|---|---|---|---|---|---|
| Anatoly Karpov (7) | Russia | 1951 | 1975 | 24 | Stamp collector, humanitarian activist |
| Garry Kasparov (6) | Azerbaijan | 1963 | 1985 | 22 | Political activist, writer |
| Alexander Khalifman | Russia | 1966 | 1999 | 33 | Chess coach |
| Vladimir Kramnik (3) | Russia | 1975 | 2000 | 25 | Foreign languages student |
| Viswanathan Anand (5) | India | 1969 | 2000 | 31 | Bachelor of commerce |
| Ruslan Ponomariov | Ukraine | 1983 | 2002 | 19 | Degree in law |
| Rustam Kasimdzhanov | Uzbekistan | 1979 | 2004 | 25 | Chess coach |
| Veselin Topalov | Bulgaria | 1975 | 2005 | 30 | Chess writer |
| Magnus Carlsen (4) | Norway | 1990 | 2013 | 23 | Entrepreneur |

*Note:* Number in brackets indicates number of times winning the world championship.

professional degrees between 1755 and 1960 (eight players: 31%) than between 1963 and 2013 (two players: 8%). The second point is a significant negative correlation between the individual's birthdate and the player's age when achieving the championship ($r = -0.40$, $p < 0.05$), indicating that the World Chess Championship has been won by increasingly younger individuals. The oldest individual included in Table 6.3 is Louis de La Bourdannais in 1834, at thirty-nine years of age, whereas the youngest world champion was Ruslan Ponomariov in 2002, when he was only nineteen years old. This is a difference in age of twenty years across two centuries. These two facts highlight that top-level chess has become more professionalized and competitive over the years. Elite world chess has become more demanding while requiring full-time involvement in the domain. For example, the stringent conditions at top-level competitive chess might hamper the investment of additional time and effort in 'extra-curricular' activities outside the chess domain. This has been conjectured in fact as a sign of lower intelligence on the part of modern chess masters, in contrast to their earlier counterparts, who had other complex interests and occupations apart from chess (Gobet et al., 2002). Besides, younger top-level chess players have been increasingly well prepared to confront and succeed against more experienced older players, though it is unlikely that this phenomenon is attributable to a generalized rise in overall intelligence, as suggested by the Flynn effect (Howard, 1999).

Appendix 3 shows the studies relating a psychometric measure of human intelligence with chess skill ($n = 30$). The links between intelligence and chess performance evolve throughout time and are typically characterized by individual differences in both child and adult populations. This body of knowledge is summarized for the seventeen studies carried out with children, and for the ten studies carried out with adults. There were two additional studies, using the same extensive sample from the Amsterdam Chess Test that comprised both children and adults (Blanch, García, Llaveria, et al., 2017; van der Maas & Wagenmakers, 2005), and one study that compared children and adults (Schneider et al., 1993). The recorded variables for each study were the sample size, the age range, the cognitive ability, the criteria used to estimate chess skill, the study design (experimental, correlational), the male to female ratio (M:F), and the country where the study took place. The measures of cognitive abilities were derived from psychometric tests such as the WISC III, Raven's Progressive Matrices, or the Intelligenz-Struktur-Test, or from other cognitive abilities such as processing speed, metacognitive abilities, language productivity, visual memory, abstract reasoning, and memory span. For the chess skill measure, most of the studies considered the Elo rating as the measure of chess expertise. Other measures of chess skill were derived from specific chess tests.

## 6.3 Intelligence and Chess in Children

There were seventeen studies with children (57%) with an age range from six to sixteen years. Six studies did not report the age range of the participants. The sample sizes ranged from twenty-four to 508, with a mean sample size of eighty-nine subjects (*sd* = 79). Five studies used the WISC, with three studies employing different subscales of this test, and two studies employing a general IQ test. Four studies used Raven's Progressive Matrices (RPM). Other studies used the Primary Mental Abilities and Differential Aptitude tests (PMA, DAT), non-verbal intelligence tests (Dearborn test, TONI-3), measures of metacognitive abilities, language productivity, abstract reasoning, calculation, and an unidentified IQ measure. Seven studies used the Elo rating to gauge chess skill, whereas five studies used some sort of independent test. This makes sense, because some children in their initial chess careers might lack a reliable Elo rating. Fourteen out of the seventeen studies adopted an experimental research design (82%), whereby some sort of comparison was undertaken between chess players versus non-chess players. In contrast, only three studies adopted a correlational research design (18%). All the studies reported a remarkable predominance of boys over girls. Male to female ratios ranged from 2:1 to 10:1, whereas three studies included only boys, with 180, forty-four, and 508 participants, respectively. Regarding the geographical area, there were three studies from Spain, two studies from the United States, and one study each from the United Kingdom, the Netherlands, Belgium, Romania, Uruguay, South Korea, Iran, Argentina, Cuba, Germany, Australia, and Zaire.

In the light of all the findings reported by these studies, it can be conceived that intelligence helps children to succeed in chess. More specifically, both general and visuospatial cognitive abilities are deemed to be necessary for obtaining a high level of chess skill in the group of Belgian chess players (Frydman & Lynn, 1992). In addition, the importance of visuospatial abilities in chess is also argued to be a key factor, with regard to the remarkable discrepancy in the presence of women compared to men, because women tend to score about one standard deviation below men in visuospatial abilities. Spatial and logical abilities are also considered useful for identifying chess talent in an independent study from the United States (Horgan & Morgan, 1990), with data from twenty relatively experienced chess players. This group scored significantly higher than average on the Raven's Progressive Matrices. Besides, when comparing twenty-two chess players from Cuba with twenty-two children uninvolved in chess, it was found that the chess players scored meaningfully higher in the Wisconsin Card Sorting Test (WCST), a measure of abstract thinking (Rojas Vidaurreta, 2011).

On the other hand, there is also evidence of non-significant differences in the Raven's Progressive Matrices test when comparing two groups of chess players at different levels of success in competitive chess (Hernández & Rodríguez, 2006). Moreover, in the study on the recognition of chess positions with children and adults, there were no differences in the digit span memory test when comparing expert and novice chess players (Schneider et al., 1993). This sort of null effect in comparing chess experts with chess novices was also observed in an interesting study with US chess players that addressed the language productivity derived from an interview comprising three main tasks: general conversation, chess conversation, and chess explanation (Nippold, 2009). The participating children in this study were classified into eighteen novices and fourteen experts by a fifty-year-old male US Chess Federation chess master. Both groups of players produced similar amounts of language and spoke with higher levels of syntactic complexity during the chess explanation task, according to several measures of language productivity: T-units, mazes, mean length of T-units, clausal density, nominal clause use, relative clause use, and adverbial clause use. This finding somewhat contradicted the expectation of experts outperforming novices because of their greater chess knowledge and experience.

Two correlational studies contrasted the joint influence of age, gender, chess experience, practice, motivation, chess enjoyment, and measures of IQ regarding their influence on chess skill. The first study (Bilalić, McLeod, & Gobet, 2007a) used an IQ measure derived from four subscales of the WISC III (vocabulary, block design, symbol search, and digit span). This measure was positively albeit moderately related to chess skill. An elite chess subsample ($n$ = 23) scored meaningfully higher than the rest of the group in the IQ measures, however. In addition, it turned out that the IQ measure related negatively to chess skill, suggesting that children scoring higher in the IQ measure had a lower level of chess skill, and highlighting the lack of a clear influence of cognitive ability on chess skill. Another remarkable finding was that, within this elite subsample, children with higher IQ scores practised chess to a lesser extent than children with lower IQ scores, rendering practice the strongest predictor of chess skill. The overall findings of the study conclude that a combination of factors such as practice, experience, age, and gender were the most likely to impinge on chess skill. The second study, with twenty-two children who were newcomers to the field of chess, used the same WISC III vocabulary, block design, symbol search, and digit span subscales as in the study by Merim Bilalić et al., together with an overall IQ measure (de Bruin et al., 2014). The main findings in this latter study indicated a strong influence by the IQ measure on chess skill, whereas practice had a somewhat lower, albeit meaningful, effect on chess skill. Moreover, motivation had a notable indirect effect on chess skill, through practice. Children who reported higher motivational levels were those reporting higher dedication to playing chess while being more involved in practice activities. The findings from these two latter studies support two main conclusions. First, cognitive ability is particularly influential for chess performance at the earlier stages of chess learning. In contrast, practice is likely to be particularly influential at progressively later stages of chess learning. Second, apart from cognitive ability, practice, age, gender, and motivation are plausible determinants of individual differences in chess skill.

The impression that chess bears several desirable properties thought to stimulate children's cognitive development is quite popular and prevalent worldwide. There is a field of research that has addressed this topic with enthusiasm. For example, when comparing school-age children who played chess with school-age children who did not play chess, studies from Uruguay (Grau-Pérez & Moreira, 2017), Argentina (Ramos, Arán, & Krumm, 2018), and Cuba (Rojas Vidaurreta, 2011) employed similar measures of executive functions, such as the WCST or the Stroop test. These three studies converge in arguing that chess might have a meaningful impact on the development of cognitive functioning at early ages. The implicit underlying idea within this contention is that practising chess at early life stages may be stimulative of a number of complex problem-solving abilities amenable to generalize and transfer to other circumstances.

Analogous findings have been reported in two studies with an extensive sample of Spanish chess players (Aciego, García, & Betancort, 2012, 2016). These studies compared 170 children involved in chess with sixty children involved in soccer or basketball as extra-curricular activities. Children were evaluated twice with the WISC-R, and with the Multifactor Self-Assessment Test of Child Adjustment (TAMAI), which taps personal, social, family, and school adjustment. The first assessment (pre-test) took place towards the beginning of the school year, and the second assessment (post-test) at the end of the school year, approximately nine months later. In the first study (Aciego et al., 2012), greater improvements in the chess group between both assessments were reported in several of the studied areas. Children in the chess group obtained significantly higher scores in the similarities, digits, block design, object assembly, and maze subtests of the WISC-R, and also in the personal and school adjustment TAMAI subtests. A similar group comparison design was undertaken in the second study, though the chess group was additionally split into a group focused on chess to spur mental skills and social values, and another group focused on the sporting side of chess (Aciego et al., 2016). Again, meaningful improvements were higher in the chess groups compared with the soccer or basketball group. Furthermore, for the chess group with an instructional method addressing mental skills and social values, the improvements were found in both cognitive and adjustment variables. For the chess group emphasizing the teaching of chess on the sporting side, the improvements were limited to the cognitive variables.

Further similar studies have yielded debatable results. A research design with twenty Romanian chess players compared with a control group of eighteen students, for example, reports that there were no differences between both groups in an IQ test, in an auditory word memory test, and in digit memory tests (Gliga & Flesner, 2014). In contrast, the chess group had a greater improvement than the control group in academic achievement in mathematics and language. Another study, with a more extensive sample size of 180 male students from Iran, allocated at random eighty-six students to a six-month chess course and another group of ninety-four students to a control group (Kazemi, Yektayar, & Abad, 2012). Both groups were evaluated in metacognitive abilities and mathematical problem solving. Higher meaningful improvements are reported for the group taking the chess course than for the control group, in both metacognitive abilities and mathematical problem solving, suggesting the usefulness of chess in promoting higher-order thinking skills.

Controversial findings have also been reported by studies describing chess instructional interventions aimed at improving the cognitive abilities of children with special needs. A study in South Korea investigated whether chess had beneficial effects in students at risk of academic failure. A ninety-minute chess programme was delivered once a week over three months to a randomly selected group of eighteen students and a control group following regular teaching activities (Hong & Bart, 2007). No cognitive effects measured by the Raven's Progressive Matrices test and a test of non-verbal intelligence were detected as a consequence of the chess instructional programme. The observed changes in the experimental group were very similar to those observed in the control group. These outcomes were partly attributed to the limited time devoted to the chess intervention, however, and to the failure of the participants to reach a minimum level of chess skill. In contrast, another study from Germany analysed whether a chess instruction intervention extending throughout a whole academic year might improve the calculation and concentration abilities of students in the low range (70 to 85) of IQ scores (Scholz et al., 2008). This study assigned a random class of thirty-one students to the chess instruction, consisting of one weekly hour of chess lessons (experimental group), while twenty-two students followed their regular classes. There was a meaningful improvement in simple addition, counting, and calculation abilities, but not in concentration abilities. The study emphasized the learning value of chess for children with learning disabilities, and that skills derived from chess lessons were somehow transferred to improving basic mathematical skills.

The conclusions from the studies arguing for meaningful improvements in cognitive abilities through chess training appear extremely positive in terms of underlining the potential benefits of chess. The findings should be taken with a substantial amount of caution, however. Several issues have been raised that suggest the need for considerable scepticism with regard to most findings from this field of research (Bart, 2014; Gobet & Campitelli, 2006; Sala, Foley, &

Gobet, 2017; Sala & Gobet, 2016). For example, a meta-analysis with twenty-four studies addresses whether there were transfer effects from chess instruction to academic and cognitive skills. None of the examined studies used a research design covering a combination of critical aspects such as the inclusion of pre-test and a post-test analysis, the random allotment of participants to different experimental conditions, or the inclusion of placebo and active-control groups (Sala & Gobet, 2016).

An additional problem with studies reporting significant improvements in cognitive abilities when comparing chess players with other groups of students is that concerning the issue of statistical power. Most of these studies adopting a comparative approach base their conclusions on findings from tests contrasting binary hypotheses: a null hypothesis stating no differences between groups against an alternative hypothesis stating differences between groups. The power of such statistical contrasts is the probability of rejecting the null hypothesis when the alternative hypothesis is true. Thus, high levels of power are desirable in a typical *t*-test that compares whether one group of chess players improves in cognitive abilities from a pre-test to a post-test, or whether a group of chess players improves in cognitive abilities compared with a group of students unacquainted with chess. Unfortunately, these kinds of statistical contrasts might be very underpowered, particularly when relying on small sample sizes, which were the most commonly used in these designs, or when the studied groups have very different sample sizes (Cohen, 1988). Ideally, these kinds of comparative designs ought to render a power analysis prior to the actual data collection.

A study applying data analyses techniques that were more sophisticated while incorporating a larger sample size yields contradictory findings to those highlighting the benefits of chess for the cognitive and academic development of schoolchildren. This study was undertaken with over 500 Australian students from grades 6 to 12, including sixty-four regular chess players. The study was not limited to a *t*-test comparison. More specifically, it used item response theory and hierarchical linear modelling to evaluate the effect of playing chess on individual differences in scientific thinking in the framework of the Australian Schools Science Competition (Thompson, 2003). This study found no evidence for the hypothesis that playing chess leads to improved scientific thinking. Instead, these findings from Australia indicate that grade level and IQ are far stronger predictors of scientific thinking than playing chess, explaining over 50% of the variability in science scholastic performance. The problem of whether chess has benefits for educational attainment and cognitive development is further explored in Chapter 10, which summarizes the application of several chess instructional interventions aimed at improving academic achievement and other desirable attributes for children.

## 6.4 Intelligence and Chess in Adults

There have been ten studies with adults (33%) with an age range from fifteen to eighty-one years. Three studies did not report the age range of the participants. The sample sizes used in the studies with adults ranged from twenty-nine to ninety, with a mean of forty-four subjects (*sd* = 19). Four studies used the Raven's Progressive Matrices (RPMs), two studies used the Intelligenz-Struktur-Test, two studies used measures of visual memory, and one study used the Berlin Structural Model of Intelligence, the WAIS, and a measure of processing speed. Unlike the research works with children, all studies with adults measured chess skill with the Elo rating. Like the studies with children, however, most studies with adults employed an experimental research design (eight: 80%), while only two studies adopted a correlational research design (20%). Similarly, the same predominance of males was also observed here, with male to female ratios ranging from 2:1 to 33:1, and with two studies including only male participants. Concerning the geographical area, there were three studies from Germany, two studies from Austria and China, and one study each from Switzerland, the United States, and the United Kingdom.

Two brain-imaging studies used the RPMs and a measures of visuospatial abilities as a control variable (Duan et al., 2012; Hanggi et al., 2014). The study by Duan and collaborators compared fifteen master-level chess players with fifteen novice-level chess players (*n* = 15). Similarly, the study by Hanggi and collaborators compared a group of twenty expert players with a group of twenty men unfamiliar with chess. Both studies report that the two groups under scrutiny scored very similarly in the psychometric measures of cognitive ability, without bearing statistically significant differences. Both studies also report remarkable brain anatomical differences, however. Expert players compared with novice players had smaller caudate nuclei, and a more extensive default brain network (Duan et al., 2012). In addition, chess players compared with controls had lower grey matter volume in the occipito-parietal junction, but very similar volumes in the caudate nucleus (Hanggi et al., 2014). These two studies attempt to explain these findings in terms of a synaptic pruning mechanism thought to elicit a more efficient integration of brain functioning (see Chapter 5).

Very similar findings regarding intelligence measures have also been reported when comparing a group of twenty-five German chess players with a control group of twenty-five non-chess players (*n* = 25) matched on age and educational level (Unterrainer et al., 2006, 2011). Both groups scored similarly in fluid abilities, or in verbal and visuospatial working memory. An additional evaluation of planning abilities was conducted with the Tower of London test, a neuropsychological test gauging planning abilities, in which chess players outperformed non-chess players. This outcome was particularly manifest for more difficult problems, whereby chess players also showed longer planning and movement execution times. One important consideration derived from this work was whether motivational or strategic differences might partially explain these findings, a topic addressed in a subsequent study from the same research group with a similar paradigm (Unterrainer et al., 2011). Another group of thirty experienced chess players were compared with a control group of thirty people in Experiment 1. Eighteen months later, in Experiment 2, twenty-two of the same chess players as those participating in Experiment 1 were compared again, with nineteen controls. The Tower of London test was applied with time restrictions in Experiment 1, and without time restrictions in Experiment 2. This study yielded no significant differences between chess players and non-players, irrespective of the time constraints imposed in each of the two experiments. Chess players reported a higher level of trait and state motivation across both experiments, however. On the other hand, the overall outcomes suggested that planning performance is equivalent in chess players and controls, running counter to the idea of transfer of chess planning to a different cognitive domain.

Visuospatial abilities have been routinely thought to be very relevant for chess playing, according to studies about mental imagery in blindfold chess (Saariluoma & Kalakoski, 1997; Saariluoma et al., 2004) and studies about intelligence and chess playing in children (Frydman & Lynn, 1992; Horgan & Morgan, 1990). Whether visuospatial abilities relate explicitly to chess skill in adults was studied by evaluating a group of thirty-six British chess players on a visual memory test (Waters, Gobet, & Leyden, 2002). Master-level players performed the same as non-master-level players, however, and the same as a broad normative sample of 550 US Navy recruits in this test. The performance in the visual memory test was in addition uncorrelated with chess skill as measured by the British Chess Federation rating. The possibility that the lack of correlation between the visual memory test and chess skill could be due to a restriction of range in the chess skill measure was disregarded because of the relatively wide range in these data. Alternatively, it was argued that this discrepancy with previous findings could depend on the narrower nature of the cognitive ability used (i.e., visual memory test versus WISC subscale), and on the sample background (i.e., adults versus children).

Although these findings indicate that visuospatial abilities might be relatively unimportant for the acquisition of chess skill and expertise in the long term, further evidence supports the view that there are meaningful individual differences regarding the association of other cognitive abilities with chess skill. For example, one of the very first psychometric studies about the intelligence of chess players (Doll & Mayr, 1987) reports significant differences in measures of cognitive abilities between a group of twenty-seven expert chess players, with Elo ratings between 2220 and 2425, and a normative sample ($n = 204$). Chess players scored significantly higher than the normative sample in processing speed, information processing, and numerical abilities, but also in general intelligence as measured by the Berlin Structural Model of Intelligence (BIS) and the Culture Fair Intelligence (CFT-3) tests. The BIS comprises three main broad factors (verbal, figural, and numerical), obtained through many different tasks, and a general measure of intelligence. The CFT-3 is a measure of fluid intelligence, comprising inductive reasoning tasks based on geometric figures. On the other hand, the intelligence test scores were uncorrelated with the Elo ratings, probably because of the restriction of range in the observed measures of chess skill.

Moreover, there are remarkable individual differences in the perceptual processing ability of chess positions. The advantage of experts over novices has been attributed to higher levels of chess experience and knowledge (Kiesel et al., 2009; Reingold et al., 2001), and to the greater familiarity of experts regarding specific meaningful arrangements of chess positions (Bilalić, McLeod et al., 2009; Schneider et al., 1993). The recognition of check and threat events in several chess positions was addressed by evaluating age effects and information-processing speed with the Digit Symbol Substitution Test from the WAIS (Jastrzembski, Charness, & Vasyukova, 2006). A group of twenty-nine young chess players between seventeen and forty-four years of age was compared with a group of thirty older chess players between forty-five and eighty-one. Slower responses from older players were expected because of decrements in information-processing abilities with ageing, although a weaker effect at higher levels of expertise was hypothesized. Nevertheless, the chess skill level did not ameliorate age-related effects on the speed of detection of checks and threats, suggesting that knowledge activation processes tend to become slower with age even for expert players.

The most comprehensive studies of intelligence and chess playing, however, were probably those performed with Austrian chess players (Grabner et al., 2006, 2007). These works are noteworthy because they comprise elements from the three levels of analysis and measurement in differential psychology, including a psychometric testing of traits, experimentation about cognitive processes, and psychophysiological recordings of the biological organism (see Figure 4.3). In the first study (Grabner et al., 2006), the EEG alpha-band event-related desynchronization method characterized the degree and topographical distribution of cortical activation of a substantial sample with forty-seven chess players. The study tested the neural efficiency hypothesis, which predicts that more intelligent people tend to display lower brain activation than less intelligent people when challenged with a cognitively demanding task (Neubauer & Fink, 2009a). The psychometric measures of intelligence were derived from the Intelligenz-Struktur-Test 2000 R, comprising measures of verbal, numerical, figural, and general cognitive abilities. There were three experimental chess-related tasks. First, the speed task consisted of determining the presence of certain chess pieces on the chessboard in a fast and accurate way. Second, the memory task stipulated the memorization of chess positions presented briefly for ten seconds. Third, the reasoning task involved solving a checkmate or chess-planning problem. The study applied an extended expert–novice paradigm, splitting the sample into four groups: (1) a lower-IQ group ($n = 23$; M = 106 [*Sd* = 9]); (2) a higher-IQ group ($n = 24$; M = 129 [*Sd* = 6]); (3) a lower-Elo-rating group ($n = 24$; M = 1717 [*Sd* = 164]); and (4) a higher-Elo-rating group ($n = 23$; M = 2076 [*Sd* = 105]). The findings in the study are aligned with the NEH. Chess players with higher scores in cognitive abilities displayed more efficient brain functioning than their colleagues with lower scores in cognitive abilities did. These individual differences were particularly evident concerning the prefrontal cortex. In addition, players with a higher level of skill in the memory and reasoning tasks also had lower activation of the frontal cortices, even with higher activation of the parietal cortices. Hence, the view is that, rather than depending on domain-specific competences and knowledge, chess performance depends largely on the general efficiency of the information-processing system.

In the second study (Grabner et al., 2007), a large sample of ninety tournament players with a mean Elo rating of 1869 (*Sd* = 247) completed several psychometric measures in intelligence, personality, motivation, and emotional competences. Moreover, the study evaluated individual differences in chess attitudes and chess practice activities. When submitting all these factors to a predictive analysis, several variables contributed to the observed individual differences in chess skill. Numerical cognitive abilities, age at entering the domain, age, number of tournament games played between 2002 and 2005, control over the expression of emotions, and motivation explained a substantial 55% of the variability in the Elo chess rating. Further additional analyses of these data addressed the association of fluid and crystallized abilities (Cattell, 1963, 1987) and their relationship to the Elo rating (Grabner, 2014a). Fluid intelligence includes subscales about sentence completion, analogies, finding similarities, arithmetic problems, number series, arithmetic operators, figure selection, cube task, and matrices. Crystallized intelligence included knowledge subscales in the general domain, and in verbal, numerical, and figural domains. Several of these subscales correlated meaningfully with the Elo rating, ranging from 0.28 to 0.44 for fluid abilities, and from 0.24 to 0.45 for crystallized abilities, and with higher correlations for number series (0.44) and numerical knowledge (0.45). Overall, these findings definitely support the view that the level of chess skill depends not only on cognitive abilities but also on individual differences in other variables.

## 6.5 Summarizing Findings about Intelligence in Chess

An in-depth review comprising the extant body of research into intelligence and chess highlights two important themes. First, expert chess players have only a modestly higher level of intelligence than control groups. Second, a chess player's strength, usually measured with the Elo chess rating, relates moderately to the level of intelligence (Grabner, 2014a). This overall review encompasses a few additional points to bear in mind.

1. Intelligence is far from being the best predictor of chess performance. An important proportion of individual differences in chess performance can be accounted for by practice within the domain.
2. Intelligence can influence the development of chess expertise in two ways. First, a minimum level of intelligence is needed to accomplish a high level of chess performance. Beyond that point, individual differences in non-ability factors (i.e., personality, motivation, and concentration) determine peak performance. Second, intelligence is particularly important in the earlier stages of chess expertise development, and its influence decreases in the later stages.
3. Individual differences in visuospatial abilities relate more strongly to chess skill and chess training performance in children than in adults.
4. There are several methodological limitations in this body of research, namely a lack of larger representative samples, a range restriction in the measures of cognitive abilities, the need for a more comprehensive measurement of cognitive abilities, and the ignoring of individual differences in practice, experience, age, gender, and personality.

Much of this body of research into intelligence and chess has been meta-analytically reviewed in a more recent study focusing on the relationship between cognitive ability and chess skill (Burgoyne et al., 2016). This meta-analysis involved three main questions: (1) whether more skilled players scored higher in cognitive abilities tests than less skilled players; (2) whether this relationship varied with age; and (3) whether this relationship varied with the content of the cognitive ability measure (i.e., visuospatial, numerical, or verbal). Cognitive ability was conceptualized here in accordance with the Cattell–Horn–Carroll (CHC) psychometric model (McGrew, 2009), contemplating a general factor (*g*), and four broad factors: fluid reasoning (*Gf*), comprehension-knowledge (*Gc*), short-term memory (*Gsm*), and processing speed (*Gs*). The main findings with regard to the association of cognitive abilities with chess skill are as follows.

1. A rather moderate association, which is somewhat similar regarding the four broad factors of cognitive ability (*Gf*, *Gc*, *Gsm*, and *Gs*), though very low for the general intelligence or *g* factor.
2. The association of cognitive abilities with chess skill is stronger for unskilled or unranked samples than for samples formed by more skilled or ranked players.
3. The association of cognitive abilities with chess skill is stronger for child and youth samples than for adult samples.

4. The association of cognitive abilities with chess skill is stronger for numerical ability content, intermediate for verbal ability content, and weak for visuospatial ability content.

Altogether, the findings from this meta-analysis are rather inconclusive about the association of chess skill with cognitive ability. Despite the fact that chess skill correlates positively with the cognitive abilities of choice, a considerable amount of the variability in chess skill (mostly over 90%) goes unaccounted for by cognitive abilities. The limited amount of studies involved could be a noteworthy drawback, however, in terms of impairing the stability of the meta-analytic results (Rosenthal, 1995). For example, the rather low number of studies ($n$ = 19) stands in sharp contrast to other meta-analyses of topics involving cognitive abilities, such as those for the Flynn effect, with $n$ = 285 studies (Trahan et al., 2014), or gender differences in mathematical performance, with $n$ = 242 studies (Lindberg et al., 2010). This limitation concerning the amount of studies included in a meta-analysis has also been highlighted in another meta-analytic review with only $n$ = 7 studies comparing chess players with non-chess players (Sala, Burgoyne, et al., 2017). This study evaluates the academic selection hypothesis, postulating that the differences in cognitive abilities between experts and non-experts are due to the training opportunities involved in access to formal academic training. Because chess lacks such an effect, unlike the admission tests to several education programmes, meaningful differences in cognitive abilities between chess players and non-chess players should record the true impact of cognitive abilities in expertise. The main findings of this meta-analysis in fact suggests an advantage in the cognitive abilities of chess players over non-chess players, implying that cognitive ability is an important explanatory factor in the development of chess skill.

Hence, and taken together, the findings from this body of knowledge suggest that the evidence attempting to link cognitive abilities to chess expertise is inconclusive and, at times, contradictory. One possible reason for this outcome stems from the idea that chess performance may largely depend on the combination and synergies of individual differences in several traits or broad clusters of traits, as suggested by the PPIK theory (Figure 4.5) (Ackerman, 1996; Ackerman & Heggestad, 1997). Individual differences in ability, but also in non-ability factors such as personality and interests, may play a role in determining individual differences in chess performance. In the next section, this question is examined with part of the data from the sample used in the development of the Amsterdam Chess Test.

## 6.6 Chess Skill versus Chess Motivation in Predicting Chess Performance

Individual differences in cognitive abilities are, obviously, important for chess performance. The seminal work by de Groot puts forward some of the main areas whereby individual differences should emerge in the elicitation of chess talent (de Groot, 1965), which include but are not limited to higher scores in cognitive and learning abilities, an extensive knowledge base, and deep motivation. The PPIK provides a theoretical framework useful to address adult intellectual development (Ackerman, 1996; Ackerman & Heggestad, 1997), comprising four main clusters of traits (see Figure 4.5): intelligence as process (*Gf*), personality, interests, and intelligence as knowledge (*Gc*). On the other hand, contemporary empirical works examining the interrelationship of cognitive abilities with chess skill and chess performance, in both children and adults, find that the evidence remains elusive, resulting in calls for comprehensive studies incorporating a wider array of individual differences comprehending practice factors and non-ability traits (Bilalić, McLeod, & Gobet, 2007b; de Bruin et al., 2014; Grabner et al., 2007). For example, the findings from Austrian chess players contemplated factors belonging to domains other than intelligence, such as experience and chess practice activities, personality, emotional competences, motivation, or attitudes. These findings are in accordance with earlier approaches to explaining chess thinking, which argue for the complex interplay of emotional, motivational, and cognitive processes (Tikhomirov & Vinogradov, 1970).

In this section, chess skill and chess motivation are compared with regard to their influence on performance in three main kinds of chess problems: tactical, positional, and endgames. More specifically, the analysis looks at whether chess skill or chess motivation are more predictive of chess performance when embedded together in the same predictive model. The data for this analysis were taken from the Amsterdam Chess Test (van der Maas & Wagenmakers, 2005). The chess players were those with complete data in all measures selected for the current analyses ($n$ = 225). These players had a mean age of thirty-one years (*Sd* = 15), ranging between eleven and seventy-eight, and a mean Elo rating of 1870 points (*Sd* = 293), ranging between 1169 and 2629 points. Apart from the Elo rating, the measures were the motivation questionnaire, and the performance in the two tests from the choose-a-move subtask (see Figure 4.6). The thirty-item motivation questionnaire measures three motivation-related traits – positive fear of failure, negative fear of failure, and desire to win – even though a global motivation score was used here. Ten seconds were allotted to complete each of these items, which were answered on a five-point disagree/agree scale. The choose-a-move A and B subtasks contain forty chess problems depicted in chessboards, and including twenty tactical items, ten positional items, and ten endgame items. There was a time limit of thirty seconds to complete each item. A correct answer to an item scored one point, while wrong answers scored zero points. Higher scores in the motivation questionnaire and in the choose-a-move A and B tasks indicated higher motivational levels, and superior performance in the tactical, positional, and endgame chess-playing dimensions, respectively.

The influence of chess skill as measured by the Elo rating and of the motivation measure on the tactical, positional, and endgame performance was evaluated with a structural equation model with latent variables – i.e., conceptual unobservable constructs that are measured with observed indicators (Bollen, 1989). This technique allows for the simultaneous specification and estimation of complex causal relationships between several variables. The specification of this model is shown at the top of Figure 6.3. There were two observed variables (chess skill and motivation), and three latent variables (tactical, positional, and endgame) measured by the two test forms of the choose-a-move subtask, A and B. Chess skill and motivation were specified as correlated, because higher Elo scores related to higher motivation scores ($r = 0.21$, $p < .001$). The model was specified and estimated with the maximum-likelihood robust method implemented with the lavaan package from the R software (R Development Core Team, 2015; Rosseel, 2012). Three different models were compared concerning the effects of chess skill and motivation on the tactical, positional, and endgame performance. Model 1 evaluated the concurrent effects of both chess skill and motivation. Model 2 evaluated the effects of chess skill only by setting the effects of motivation on tactical, positional, and endgame variables to zero. Model 3 evaluated the effects of motivation only by setting the effects of chess skill on tactical, positional, and endgame variables to zero. The effects represented with numbers are to be interpreted as regression coefficients, although, in a structural equation modelling context, are commonly referred to as beta weights (Bollen, 1989). The three models were additionally compared with a chi-square difference test ($\Delta\chi^2$). Assuming a correct model specification, a significant difference in comparing two models ($p < 0.05$) would support the most parsimonious model with the lower amount of degrees of freedom (Yuan & Bentler, 2004).

The main findings are shown at the bottom of Figure 6.3. Model 1 shows that the effects of chess skill were highly significant on each kind of performance: tactical (0.79), positional (0.85), and endgame (0.89). In contrast, the effects of motivation were significant only with regard to tactical performance (0.18, $p < 0.001$), albeit with a much lower magnitude. The coefficients of determination ($R^2$) indicate that the explained variability in each kind of performance was large: 71% for tactical performance, 75% for positional performance, and the largest – 81% – of the accounted variance for the endgame performance. The model fit indices were good (Hu & Bentler, 1999), suggesting that the specified model did indeed represent the observed data well (CFI = 0.991, TLI = 0.979, RMSEA = 0.061, AIC = 4,014). The Model 2 findings show that, when the effects of motivation on the tactical, positional, and endgame latent variables were set to zero, there was a meaningful worsening of model fit when compared with Model 1, with a significant chi-square difference: $\Delta\chi^2[3] = 15.18$, $p = 0.0017$. Model 2 shows

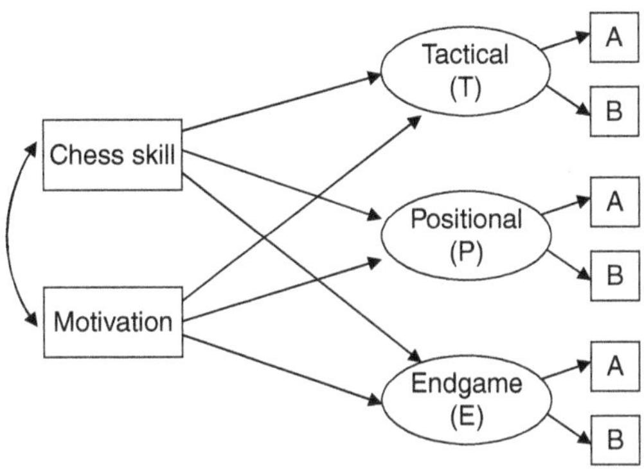


| | Model 1 | | | Model 2 | | | Model 3 | | |
|---|---|---|---|---|---|---|---|---|---|
| $\chi^2$ | 22.02* | | | 37.20 | | | 281.85*** | | |
| CFI | .991 | | | .980 | | | .759 | | |
| TLI | .979 | | | .962 | | | .550 | | |
| RMSEA | .061 | | | .083 | | | .286 | | |
| AIC | 4,014 | | | 4,024 | | | 4,276 | | |
| | T | P | E | T | P | E | T | P | E |
| Chess skill | .79*** | .85*** | .89*** | .83*** | .86*** | .90*** | --- | --- | --- |
| Motivation | .18* | .05 | .04 | --- | --- | --- | .34*** | .23** | .23** |
| $R^2$ | .71 | .75 | 81 | .68 | .74 | .81 | .12 | .05 | .05 |

**Figure 6.3** Structural equation model evaluating the impact of chess skill (Elo rating) and motivation, on tactical (T), positional (P), and endgame (E) chess performance; observed variables are represented with squares, latent (unobserved) variables are represented with ellipses; one-headed arrows represent causal links, the two-headed arrow a correlation; there were twelve degrees of freedom for Model 1, and fifteen degrees of freedom for Models 2 and 3 (CFI = comparative fit index; TLI = Tucker–Lewis index; RMSEA = root mean squared error of approximation; AIC = Akaike information criterion)
*Notes*: * $p < 0.05$; ** $p < 0.01$; *** $p < 0.001$.

that the fit indices were still acceptable (CFI = 0.980, TLI = 0.962, RMSEA = 0.083, AIC = 4,024), that chess skill had a meaningful impact on the tactical, positional, and endgame latent variables, and that chess skill explained a considerable amount of variance in the three latent variables. The Model 3 findings show that, when the effects of chess skill on the tactical, positional, and endgame latent variables were set to zero, there was an even stronger deterioration in model fit when compared with Model 1, with a highly significant chi-square difference: $\Delta\chi^2[3] = 259.83$, p <

0.0001. Model 3 shows that the fit indices were unacceptable (CFI = 0.759, TLI = 0.550, RMSEA = 0.286, AIC = 4,276). Furthermore, even though motivation had a meaningful impact on the tactical, positional, and endgame latent variables, motivation explained a rather poor amount of variance in the three latent variables.

These findings suggest that Model 1 was the best representation of the observed data for explaining the observed individual differences in the three dimensions of chess performance. Hence, taking Model 1 as the best explanation of the observed data supports the view that chess skill is the strongest predictor of chess performance as a whole, whereas motivation exerts only a modest influence in the more specific tactical performance dimension. Furthermore, chess skill effects and explained variance ($R^2$) are progressively higher when moving from the tactical to the endgame performance latent variables (Models 1 and 2). In contrast, motivation effects and explained variance are progressively lower when moving from the tactical to the endgame performance (Models 1 and 3).

These differences of both chess skill and motivation as related to the three dimensions of performance can be explained in terms of the characteristics of each aspect of the game, calibrated by the respective items of the choose-a-move test (Dvoretsky & Yusupov, 1996; Kotov, 1971). First, tactical items consist of a single and sometimes unexpected surprising move. In addition to the fact that these items might be somehow easier for average-skilled players, they might also exert a stronger motivation to find the specific single move leading to the solution. Conversely, the positional items involve detecting more intricate and finer points that characterize a given chess position. Positional thinking may involve a more elaborated knowledge of concepts such as open lines, pawn structures, weak and key squares, weak colour complexes, or space and centre, while being more challenging to average-skilled players. Similarly, the endgame requires a highly refined technique and a deep study of the diversity of schemes usually arising at this stage of the game (Kotov, 1971). A typical example of a very specific scheme in the endgame is a rook against a rook plus a pawn. Another example is an ending with opposite-coloured bishops, which increases the chances of the game outcome being a draw. In sum, finding a solution within a limited time of thirty seconds might be much harder and less motivating for average than for skilled players, particularly regarding positional and endgame items.

These findings, as a whole, suggest that individual differences in a non-ability trait such as motivation might additionally influence chess performance. In accordance with the PPIK theory about adult intellectual development (Ackerman, 1996; Ackerman & Heggestad, 1997), individual differences in personality are likely to influence the structure and development of chess skill.

The next chapter addresses personality, a classical broad dimension and another of the central objects of study, together with intelligence, in the discipline of differential psychology. Although the association of intelligence with chess has probably been examined more deeply and extensively than that of personality, there are also some intriguing studies that have considered how individual differences in personality relate to chess skill.

# 7

# Personality

Personality is another of the nuclear domains in differential psychology (see Figure 4.2). Personality is an abstract and broad concept that comprises a combination of constitution, character, temperament, and – for some authors – cognitive traits. Initial theoretical approaches to personality relied on discrete typologies. A few behavioural types comprise a conglomerate of similarities and consistencies in behaviour. Some examples of personality typologies include the ancient four humours stemming from Greek medicine, the constitutional typologies from Kretschmer and Sheldon, and the character typology from Heymans (Andrés-Pueyo, 1997). Nonetheless, contemporary theories of personality focusing on individual differences adopt an approach based on traits generally measured with personality tests. Traits are the dispositional attributes underlying behaviours that are consistent and manifest across different situations, and tend to show stability over time. Personality tests are generally well elaborated, and reliable and valid, with high discriminative properties, and with available appropriate and comprehensive statistical norms. There are three main kinds of personality tests.

1. Questionnaires. These instruments comprise an assortment of items formulated in terms of questions, lists of adjectives, or propositions that relate to individual behaviours, feelings, or belief systems. People are asked to endorse each of the items in accordance with a response scale. Personality questionnaires are easy and relatively cheap to complete, and can be used either individually or collectively. Most personality theories and models based on the analysis of traits use this kind of instrument.
2. Projective. Tests based on the projective approach use a number of pictures that are presented to the individual to describe what he or she perceives. Personality is assessed in line with these perceptions. Projective tests are used from an ideographic approach and are mostly used in clinical contexts. Psychometric criteria such as reliability and validity are not usually applied to these kinds of tests.
3. Objective. An objective test is an instrument whereby it is possible to score in an objective way the responses of the individual while he or she is unable to suspect the aims of the evaluation. Some examples of these tests are

psychophysiological assessments, projective tests that are scored objectively, and even personality questionnaires.

Because of the robust predictive ability of personality traits regarding several notable behaviours, human personality has been studied extensively from different approaches and across different cultures. Nevertheless, it has been suggested that the cross-cultural measurement of personality is polarized along four main axes: trait/situation, universalism/relativism, quantitative/qualitative, and anthropology/psychology. None of these instances has provided a clear, unequivocal, and consistent body of evidence, however. For instance, universalism focuses on similarities, whereas relativism focuses on differences. In this sense, collaborative interdisciplinary research efforts have been put forward as a promising avenue for integrating both universals and cultural variations in human behaviour (Marsella et al., 2000; Mayer, 2005; Poortinga & van Hemert, 2001). Several studies suggest that the five-factor personality trait structure is universal, however. The findings from an extensive study with a representative sample ($n = 7{,}134$) support the view that the five-factor model (FFM) of personality replicates robustly across highly diverse cultures from five distinct language families (McCrae & Costa, 1997). Moreover, another comprehensive study across twenty-six cultures ($n = 23{,}031$) finds support for similar levels of personality traits, with factor scores being related to culture level variables (McCrae, 2001). A cross-national sample from Canada, Germany, and Japan with 1,209 monozygotic twins and 701 dizygotic twins finds in addition supportive evidence for a consistent biological basis for the five-factor model of personality (Yamagata et al., 2006).

Because a substantial 40% of the observed variability in personality is attributable to genetic influences (Vukasovic & Bratko, 2015), personality traits interrelate closely with major goals in life (Bleidorn et al., 2010). There are noteworthy individual differences in personality in the course of development, with genetic and non-shared environmental factors influencing the stabilization of personality traits over time (Briley & Tucker-Drob, 2014; Hopwood et al., 2011). Longitudinal findings from a forty-five-year follow-up study highlight the fact that the big five personality traits correlate with other life course variables, such as global adult adjustment, career success, social relationships, mental health, substance abuse, and political attitudes (Soldz & Vaillant, 1999). Personality traits are robust predictors of social relationships, educational and work performance, and health promotion (Caspi et al., 2005); they show consistent stable associations with psychological well-being during adult personality development (McCrae, 2002), albeit with age differences across different cultures (McCrae et al., 1999). Personality traits predict, for instance, critical life outcomes such as mortality, divorce, and occupational attainment (Roberts et al., 2012). For example, the traits that stimulate the selection of people into specific work experiences are the same traits that change as the likely response to these specific work experiences (Roberts, Caspi, & Moffit, 2003). The big five personality traits predict major life events (selection effects), but they change in turn as a reaction (socialization effects) to the experience of these events (Specht, Egloff, & Schmukle, 2011).

Large numbers of meta-analytic studies support the existence of a sound link between personality and major social outcomes. For example, some meta-analyses have focused on work-related topics, such as job performance (Barrick & Mount, 1991), work safety (Beus, Dhanani, & McCord, 2015), job satisfaction (Judge, Heller, & Mount, 2002), and job burnout (Swider & Zimmerman, 2010). Other meta-analyses have focused on health-related issues such as elevated blood pressure (Jorgensen et al., 1996), subjective well-being (DeNeve & Cooper, 1998), and anxiety, depressive, and substance use disorders (Kotov et al., 2010). Moreover, there are meta-analytic studies addressing general social adjustment themes, such as aggressive behaviour (Bettencourt et al., 2006), academic performance (Poropat, 2009), and parenting (Prinzie et al., 2009).

## 7.1 Approaches to the Study of Personality

As with intelligence, abstract latent personality traits are inferred from the analysis of behaviour to describe individual differences, with a variety of theoretical approaches and empirical methods (Pervin, 1990; Uher, 2008). For instance, there are computational approaches to the empirical analysis of personality traits that have been based on neural network applications and that have been mostly based on the reinforcement sensitivity theory of personality (Read & Miller, 2002; Read et al., 2010). The definition and architecture of personality have relied on data from questionnaires submitted to the statistical technique known as factor analysis, however. This technique aims to empirically derive a few comprehensible factors considered to be sufficiently representative of a myriad of behaviours (Boyle, Stankov, & Cattell, 1995). Factorial personality models attempt to describe the basic dimensions of personality, which are useful for predicting behaviour in daily life, while also being stipulated as reference models regarding personality disorders.

There are two main approaches that can be ascertained within the factorial personality models: lexical and biological. Factorial personality models within the lexical approach deal with the description of the structure of personality in terms of dimensions or factors to determine a comprehensible and acceptable nomenclature for the scientific community. Joy Paul Guilford's model is probably one of the first attempts in this line of research. This model proposes a hierarchical five-level description of personality that places four main factors towards the upper levels of the hierarchy (Guilford, 1975): social activity (SA), introversion/extraversion (IE), emotional stability (E), and paranoid disposition

(Pa). Cattell's model is another conceptualization within the lexical approach, stemming from the premise that natural language underpins the description of human personality (Cattell & Kline, 1977). Raymond Cattell's model builds on previous lexical work from Gordon Allport and Henry Odbert, who extracted 17,953 unique terms useful for describing personality and behaviour (Allport & Odbert, 1936). The factor structure obtained by Cattell includes sixteen primary personality factors, with subsequent developments giving rise to five broad personality factors (Cattell, Eber, & Tatsuoka, 1970): introversion/extraversion, low anxiety/high anxiety, receptivity/tough-mindedness, accommodation/independence, and lack of restraint/self-control. This model is one of the more complex descriptions of personality, with high impact in applied fields, such as health, work, and education.

In the forty years after the exhaustive work by Cattell, the studies addressing the organization of human personality have been fairly successful in replicating a five-factor structure. For example, work aimed at evaluating the effectiveness of air force officers showed a five-factor structure comprising surgency (heightened positive affect), agreeableness, dependability, emotional stability, and culture (Tupes & Christal, 1992). Further research aimed at the construction of systematic structured taxonomies of adjectives derived from natural language (Goldberg, 1990). The scores in the scales elaborated by Goldberg correlate well with the NEO personality inventory (NEO-PI), a personality questionnaire that raises an orthogonal five-factor solution and that has been extensively replicated in subsequent developments of this instrument (Costa & McCrae, 1985, 1992). The NEO-PI comprises five factors (neuroticism, extraversion, openness to experience, agreeableness, and conscientiousness), together with six narrower facets per factor. Table 7.1 displays the main structure of the NEO-PI. Neuroticism relates to emotional stability, anxiety, hostility, and impulsiveness. Extraversion relates to the amount and intensity

Table 7.1 *Factors (in boldface) and facets of the five-factor model (FFM) measured with the NEO-PI-R instrument*

| |
|---|
| **Neuroticism (N):** anxiety (N1), anger/hostility (N2), depression (N3), self-consciousness (N4), impulsiveness (N5), vulnerability (N6) |
| **Extraversion (E):** warmth (E1), gregariousness (E2), assertiveness (E3), activity (E4), excitement seeking (E5), positive emotions (E6) |
| **Openness to experience (O):** fantasy (O1), aesthetics (O2), feelings (O3), actions (O4), ideas (O5), values (O6) |
| **Agreeableness (A):** trust (A1), straightforwardness (A2), altruism (A3), compliance (A4), modesty (A5), tender-mindedness (A6) |
| **Conscientiousness (C):** competence (C1), order (C2), dutifulness (C3), achievement striving (C4), self-discipline (C5), deliberation (C6) |

of social bonds, energy, and sensation seeking. Openness to experience is a controversial factor, including traits related to intellect and creativity. Agreeableness includes traits focused on interpersonal relationships, such as empathy. Finally, the conscientiousness factor includes traits with a focus on major aims and goals in life, emphasizing adherence to social norms. This is in fact the dominant model at present regarding basic and applied research into human personality. Furthermore, the model has been successfully replicated across different cultures and languages, while being capable of integrating a comprehensive assortment of personality constructs. More importantly, it has facilitated the communication and sharing of findings across different researchers from diverse disciplines.

On the other hand, factorial personality models within the biological approach attribute individual differences in behaviour to psychophysiological mechanisms. In contrast to models from the lexical approach focused on describing the structure of personality, biological models aim to explain behaviour in causal terms. The three-dimensional psychoticism–extraversion–neuroticism (PEN) model from Hans Eysenck (see Figure 7.1) has probably been the most influential within this approach (Eysenck, 1952). Typical descriptors for psychoticism (aggressive, cold, egocentric, impersonal, impulsive, antisocial, tough), extraversion (sociable, lively, active, sensation seeker, assertive, dominant, adventurous), and neuroticism (anxious, depressed, guiltiness, tense, irrational, shy, emotional) provide the basic pattern of behaviours captured by each broad dimension. Psychoticism relates to androgen hormones, with the fight/flight system, the mechanisms of fear, and neurotransmitters such as serotonin, noradrenaline, and dopamine (Davidson, Putnam, & Larson, 2000). Excitatory and inhibitory mechanisms explain the extraversion dimension. Extraverted people are characterized by weak excitatory potentials and fast inhibitory

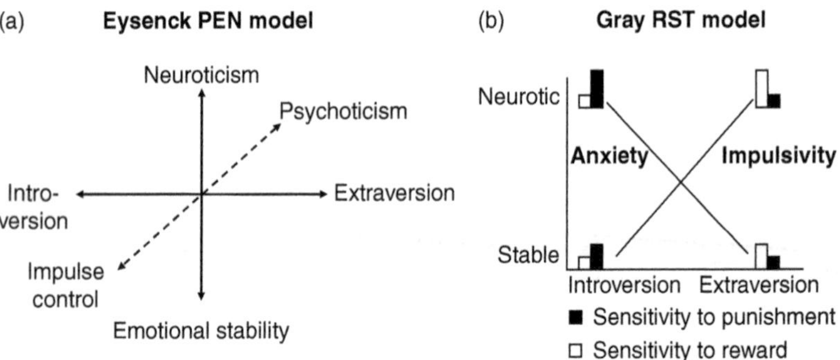


**Figure 7.1** Eysenck's PEN (psychoticism–extraversion–neuroticism) and Gray's RST (reinforcement sensitivity theory of personality) models of personality

processes, whereas, in contrast, introverted people are characterized by strong excitatory potentials and slow inhibitory processes. With inhibition contributing to decrements in performance, extraverts would have higher decrements in performance in long-lasting tasks. Moreover, extraverts have lower activity levels than introverts in the ascending reticular activating system (ARAS), a structure located in the brainstem. Individual differences in these activity levels influence the preference of extraverts for activities with higher stimulation, and the preference of introverts for activities with lower stimulation. Finally, the PEN theory stipulates that the neuroticism dimension governs the balance between the sympathetic and parasympathetic nervous system. People high in neuroticism show fast changes between the functions executed by both systems, whereas, in contrast, people low in neuroticism show slow changes between the functions executed by both systems (Eysenck, 1994; Juan-Espinosa & García Rodríguez, 2004).

Jeffrey Gray's reinforcement sensitivity theory (RST) of personality had its origins in studies with non-human animals. Although it is similar to the PEN model (see Figure 7.1), the RST postulates anxiety and impulsivity as the two broader core personality dimensions, with anxiety occupying the quadrant formed by introversion and neuroticism, and impulsivity occupying the quadrant formed by extraversion and neuroticism in the PEN model (Gray, 1987). Predictions of the RST relate to environmental signals, either of punishment or of reward. Anxiety associates with a higher sensitivity to punishment (low extraversion, high neuroticism). In contrast, impulsivity associates with a higher sensitivity to reward (high extraversion, high neuroticism). Moreover, the RST model postulates three kinds of systematic emotional responses. The behavioural approach system (BAS) is sensitive to signals of unconditioned and conditioned reward signals, forming the basis for impulsivity and extraversion. The behavioural inhibition system (BIS) is sensitive to signals of punishment, and it forms the basis for anxiety and neuroticism. The fight–flight–freeze system (FFFS) is also sensitive to conditioned and unconditioned signals of punishment, and forms the basis for fear (Corr, 2004; Smillie, 2008). There is substantial evidence for these neural underpinnings of personality. A comprehensive account from fMRI findings with healthy people has revealed an intricate complex of robust relationships between cortical and subcortical brain structures, with the main factors underpinning the RST (Kennis, Rademaker, & Geuze, 2013).

Marvin Zuckerman has elaborated another influential personality model from the biological approach. Zuckerman's model highlights a very specific personality dimension, sensation seeking (Zuckerman, 2005), representing an interest in internal sensations that are novel, varied, and complex, together with the will to become involved in risky experiences of different kinds (i.e., physical, social, or legal). This model maintains that people with a lower level of monoamine oxidases (MAOs), a family of enzymes involved in the deactivation of neurotransmitters, have higher levels of sensation seeking. This personality model is also termed the alternative five because it suggests five broad personality dimensions: impulsive sensation seeking, neuroticism/anxiety, aggression/hostility, activity, and sociability (Zuckerman, Kuhlman, Joireman, Teta, & Kraft, 1993).

Other conceptualizations of personality are embedded in evolutionary perspectives. For example, six affective systems closely linked to subcortical brain structures are thought to underlie human and other mammalian emotional states. These systems are play, seeking, care, fear, anger, and sadness, and have been suggested as a necessary step towards refining the big five personality model (Davis & Panksepp, 2011). On the other hand, evolutionary psychology has been proposed as a useful general framework for addressing personality and individual differences (Buss, 1991, 2009). In this view, several approaches to the study of personality have focused on individual differences, whereas there has also been a growing interest in cultural universals. This diversity of approaches to the study of personality might imply antagonistic research aims and methods. An integration of these different approaches has repeatedly been suggested, however, because of the high degree of complementarity (Church, 2001, 2010).

Compared with intelligence, individual differences in personality in relation to chess have been much less scrutinized. The research in this latter area has been mainly focused on three main questions. First, earlier works about the psychology of chess set forth the issue of whether individual differences in personality impinge on the playing style of chess players. Second, later works using some of the available personality models and personality questionnaires have enquired into whether the personality of chess players differs across different levels of chess skill or from the personality of the general population. Third, these later works have addressed the question of what the most predictive personality factors of chess skill might be.

## 7.2 Personality and Chess-Playing Style

The temperament of chess players was said to determine their playing style and playing habits, either towards winning or losing (Cleveland, 1907). A player's style has been defined as the judgement and decision-making fashion applied to chess playing, which would depend on individual differences in thinking, concentration, emotional state, and character (Krogius, 1976). This latter book about the psychology of chess, written by a chess grandmaster and psychologist, suggests either combinative or positional playing styles as the most common among chess players. Combinative players would be those with greater creativity and imagination abilities, keen to execute deep and very specific calculations, and conducting complex combinations including risky piece sacrificing. Positional players, in contrast, would be those with a stronger tendency to robust strategical play, relying on well-established and very general chess-playing principles.

There are several other classifications concerning chess-playing style. For example, a description of training methods concerning the opening part of the game proposes four main chess-playing styles – ambitious, conformist, robust, and risky – the four of them leading to different configurations, advantages, and drawbacks in the initial stage of the game (Peralta & de Dovitiis, 2007). Also related to the opening stage, two main chess-playing styles have been distinguished: attacker and defender. Players prone to either attacking or defending might tend to use different chess openings, broadly termed as open or closed openings, respectively. Attackers would tend to begin with the move e4, whereas defenders would tend to begin with a move such as d4, c4, Kf3, or g3 (Benko & Hochberg, 1991).

Some of these reflections, based on an extensive body of background knowledge, might lead us to think that playing style could well relate to individual differences in personality. Whether there may be a correlation between individual differences in personality with specific playing styles in chess has, with reason, been questioned, however. Chess players with steady and orderly ways of life might play with a violent and somehow disordered style, whereas some famous chess players with more undisciplined ways of life have in fact played with an overwhelming logic and an ordered style (Benko & Hochberg, 1991). On the other hand, when considering the relevant role of attention in chess, it has been argued that individual differences in knowledge and experience, aesthetic preferences, and character would exert considerable influence on the direction of attention (Krogius, 1976). Krogius also highlights other psychological attributes with an influence on attention that might underlie individual differences in personality, such as willpower, or the self-regulation of changing moods and emotions. At the same time, he also emphasizes that the character of a chess player determines his or her playing style, and that chess playing reflects personality to a large extent.

Attention in chess might also be influenced by individual differences in one's opponent's personality during a chess game. Modifying the focus of attention might induce changes in affective states during chess playing (Freund & Keil, 2013). For instance, the ability to conceal certain kinds of emotions might influence the final outcome of a chess game (Grabner et al., 2007; Llaveria et al., 2016). Other authors have also underlined the importance of being acquainted with the personality of one's potential opponents. One of the most eminent books about chess training emphasizes the need to have a deep knowledge of oneself and also of the opponent, not only as a chess player but also as a person (Kotov, 1971). For example, a chess-training strategy should include delineating an opponent's characteristics, including his or her dominant playing style (combinatorial or positional), originality and creativity, accuracy in the calculation of variants, objectivity, self-confidence, risk-taking, or the tendency to run into time troubles (Samarian, 2008). Because both opponents know in advance the chess strength of their respective opponent, chess players may choose to adopt different strategies depending on the chess strength of the opponent rather than on the opponent's personality. This topic has been addressed in an interesting study examining over 1 million internet blitz chess games (Fernandez-Slezak & Sigman, 2012). A given player confronted either stronger or weaker players in all the games. Consequently, this study analysed two strategies: a promotion mode and a prevention mode. In the promotion mode, chess players sought to approach a desired state (winning). In the prevention mode, chess players sought to avoid an undesired state (losing). Players changed their strategy by adopting a prevention mode when playing stronger opponents, which implied playing more slowly and more accurately. This strategy was deemed to be imperfect, however, because the excess of thinking time did not compensate for any subsequent game advantage.

## 7.3 Personality Factors Studied with Chess Players

In delving into the personality of expert chess players in relation to their chess achievements, de Groot argues that success in chess does not demand exceptional intelligence, but that it does depend on favourable educational opportunities (de Groot, 1965). Moreover, when putting forward the traits in which individual differences would be most likely to emerge concerning the structure and development of chess talent, he points to a deep motivational disposition, and the ability to integrate thinking with playing and fighting. Nevertheless, de Groot also suggests that the personality factors important for chess talent remain largely unknown. Further contemporary research into personality and chess have looked at whether the personality factors of chess players differ from those observed in the general population, and whether there is a characteristic personality pattern in chess players. Unfortunately, the studies addressing the personality of chess players that could shed light on these questions are rather scarce.

Appendix 4 summarizes eleven studies from 1985 to 2016. As far as is known, these are all the available studies to date addressing individual differences in the personality of chess players. Compared with the studies about intelligence and chess, there are some noticeably distinctive features regarding the studies about personality and chess. Unlike the studies about chess and intelligence, whereby child and adult samples were more balanced, only four studies used child samples when studying personality factors (36%). Personality may be somehow more difficult to evaluate with children than with adults due to methodological, logistical, and ethical reasons. Moreover, the studies about personality comprised higher numbers of participants than the studies about intelligence, with sample sizes ranging from forty to 479 participants. One of the most remarkable issues is that most of the studies used different personality instruments, such as the Minnesota Multiphasic

Personality Inventory (MMPI), the Big Five Questionnaire for Children (BFQ-C), the NEO Five-Factor Inventory (NEO-FFI), the Sensation-Seeking Scale (SSS), the Myers–Briggs Type Indicator (MBTI), the Eysenck Personality Questionnaire (EPQ), and the Freiburg Personality Inventory – Revised (FPI-R). This is a limiting factor concerning the comparison of findings across studies and across personality factors. Nevertheless, the two main broad personality factors that appear in most personality models, extraversion and neuroticism, were evaluated in one way or another throughout all these studies. Furthermore, most studies adopted a mixed research design by performing some kind of group comparison (i.e., chess players versus non-chess players) together with analysing whether personality factors were correlated with other variables. Male to female ratios ranged from 2:1 to 60:0, illustrating again the remarkable underrepresentation of women in chess. Data from four studies were from Spain, two studies took place in the United States and two in Austria, and the rest of the studies were conducted in Israel, the United Kingdom, and Germany. Akin to the studies about intelligence, the studies about personality have been carried out in wealthy Western nations.

Because all the studies used different personality instruments, comparing these findings could be problematic. There are some interesting common points worth noting, however. For example, the studies comparing chess players at different levels of skill, or chess players with non-chess players, find meaningful inter-group differences in the evaluated personality factors, such as in the work with the MBTI (Kelly, 1985). The study comprised data from over 400 US chess players, both at average ($n$ = 270) and at master ($n$ = 209) skill levels. The findings from this study suggest that chess players have differentiated personality and temperamental characteristics compared with the general population, while there are also some prominent differences between average- and master-level players. A personality factor labelled as intuition was the most discriminative between average and master chess players. This factor is positively correlated with the NEO PI-R openness to experience factor, while being strongly related to chess skill. These findings appear to demonstrate that chess might be driven by intuitive rather than by logical thinking. Nevertheless, the MBTI has been widely criticized, for ignoring a fundamental personality dimension such as neuroticism, for lacking comprehensive information about all its subscales, and for its psychometric inadequacy compared with other, more theoretically driven instruments, such as the NEO-PI (Furnham, 1996).

A subsequent study by Avni and collaborators with chess players from the Israel Chess Association compared three groups of twenty individuals: highly competitive players, moderate competitive players, and non-players (Avni, Kipper, & Fox, 1987). Highly competitive players scored higher than non-players in suspiciousness, unconventional thinking, and orderliness. There were no significant differences between the three groups in neuroticism, aggression, or hostility, however. A plausible interpretation of these findings is that both unconventional thinking and orderliness are useful and compatible (i.e., adaptive) for the characteristics of the game. The study with children is particularly interesting because it addressed this same issue, albeit at an earlier developmental stage (Bilalić et al., 2007b). These unique findings from children involved in chess indicate that lower scores in agreeableness, higher scores in extraversion, and higher scores in openness to experience are associated with a higher propensity to become involved in the chess domain. One crucial question raised in this study with child chess players, however, was whether personality would interact with individual differences in other variables such as motivation and practice, an unknown that was also raised in the study by the same authors about intelligence and chess in children (Bilalić et al., 2007a).

The underpinning multifactor nature of chess skill is a recurrent theme that was addressed in one of the studies performed with chess players from Austria. There are two studies with Austrian chess players. In the EEG study regarding the neural efficiency hypothesis, the NEO Five-Factor Inventory was used as a control measure of the EEG recordings (Grabner et al., 2006). In the second study, with ninety chess players, a comparison was made as to whether distinct traits from three different broad domains were able to predict the level of chess strength as measured by the Elo rating: (1) intelligence; (2) experience and practice activities; and (3) personality, emotional competences, motivation, and attitudes (Grabner et al., 2007). For each of the studied domains, the stronger predictors of the Elo rating were numerical intelligence, the age on entering the chess club, the actual age, the number of tournament games, control over the expression of emotions, and motivation. The combination of these variables explained over 50% of the variability in the Elo chess rating. Thus, those chess players with higher levels of numerical cognitive ability, who started to play seriously earlier and had longer chess experience, having played a greater number of games, and who had higher motivation and superior control of the expression of emotions were also those with a higher level of chess skill measured by the Elo rating. None of the big five personality factors had a meaningful impact on the variability in chess skill, however. There are two main reasons that might account for these null findings. First, the influence of personality on chess skill could be indirect. Personality factors could act as a mediator variable between specific cognitive abilities – i.e., numerical, or chess practice and chess skill. Second, personality could be influential at the time of choosing to play chess at earlier stages of the lifespan. These latter kinds of choices could be supported by comparing the personalities of chess players with those of the general population. Meaningful differences could represent a plausible piece of evidence supportive of this hypothesis. There are some empirical findings that could point to that direction.

A study with eighty-two US college students analysed whether the personality trait of sensation seeking related to chess involvement (Joireman, Fick, & Anderson, 2002). Higher scorers in sensation seeking turned out to be more attracted to chess. It was suggested that these findings corroborated the view that highly competitive players have higher levels of unconventional thinking (Avni et al., 1987), and chess players who won or were about to play a chess contest have higher levels of testosterone (Mazur et al., 1992). A study implemented by our own research group used instead a well-established instrument: the EPQ (Llaveria et al., 2016). One of the main findings is that, when comparing the personality data from a group of 100 amateur chess players with the normative data of the EPQ ($n$ = 527), chess players scored significantly lower in the three dimensions of the Eysenck's PEN model: extraversion, neuroticism, and psychoticism.

Only one study has addressed the personality of elite chess players (Vollstädt-Klein et al., 2010). Their level of chess skill was characterized with the German chess rating scale (Deutsche Wertzungszahl: DWZ), which is analogous to the Elo rating. There were data from thirty male players with between 2050 and 2575 DWZ points, and thirty female players with between 1630 and 2189 DWZ points. Aside from comparing the personality patterns of chess players with the general population, this study enquired specifically whether personality factors related to chess skill and whether there were meaningful differences between males and females. The personality of male chess players did not differ from that of the general population, but female chess players scored higher than the general population in life satisfaction and achievement orientation, but lower in somatic complaints. When relating the personality factors to the chess skill of male players, extraversion and stress correlated negatively with chess skill, suggesting that stronger male players are more introverted and less stressed than weaker male players. For female players, extraversion was positively related with chess skill, whereas inhibition and aggressiveness were negatively related with chess skill, suggesting that stronger female players are more extraverted than weaker female players, and also less inhibited and aggressive.

Three studies from Spain used an experimental research design to evaluate whether socio-affective variables improved as a result of the regular practice of chess. Two of these studies used the Multifactor Self-Assessment Test of Child Adjustment (TAMAI), which evaluates personal, social, family, and school adjustment (Aciego et al., 2012, 2016). The characteristics of these studies are described in detail in Chapter 6, about intelligence. In both studies, the chess group showed higher improvements than the soccer or basketball group in the TAMAI subtests as reported by teacher's ratings, particularly in academic and personal adjustment, and also in coping capacity. Nevertheless, the authors of the study also acknowledge that the students who chose chess initially were also those better adapted to the school context compared with the students who chose soccer or basketball. The third study used the questionnaire MOLDES, which measures cognitive affective dispositions, and applied it to fifty-three chess players between ten and sixteen years of age (Hernández & Rodríguez, 2006). Individual differences in the MOLDES instrument were compared in two groups classified in accordance with chess competition outcomes: unsuccessful and successful. This report underlines the fact that players classified as unsuccessful scored higher in evasive, magic, defensive, and inoperative dispositions. In contrast, successful players scored higher in realistic, positive, and emotional regulation dispositions. Interestingly, both groups had equivalent scores in general intelligence as measured by Raven's Progressive Matrices, and in school achievement. These findings reinforce the notion of the potential influences of individual differences in motivational and emotional traits, particularly in the earlier stages of the chess career.

The next section compares the influence of several factors on chess skill with a sample of Spanish adult chess players. The predictive factors were chess knowledge and chess motivation, the personality dimensions from the PEN model – psychoticism, extraversion, and neuroticism – and measures of emotional regulation.

## 7.4 Personality, Motivation, Emotional Regulation, and Chess Knowledge

Chess skill and chess performance are likely to depend on several factors implying individual differences in ability and non-ability factors, and their interplay (Bilalić et al., 2007b; Grabner et al., 2007; Tikhomirov & Vinogradov, 1970). Success in chess and in other intellectual domains has been regarded as entailing a degree of proficiency in certain cognitive abilities and a consistent knowledge base gained through an extensive period of practice, in addition to high levels of motivation and emotional self-regulatory skills (Ackerman, 1996; Ackerman & Heggestad, 1997; Ericsson & Charness, 1994; Grabner et al., 2007; Krogius, 1976). The findings from Austrian chess players in the research programme developed by Roland Grabner and collaborators (Grabner et al., 2006, 2007) suggest that factors belonging to different domains influence chess skill. Similarly, the aims of the study by Anton Llaveria and collaborators were, in the first place, to compare the personality and emotional regulation measures of the chess players with those from the general population. Second, the study aimed to compare the association of personality, chess motivation, and emotional regulation with measures of chess knowledge in the prediction of chess skill as measured by the Elo rating. This section describes the findings of this study in more detail (Llaveria et al., 2016).

The participants were 104 players enrolled in the Catalan Chess Federation who competed regularly in chess tournaments. This group of chess players completed the verbal chess knowledge and chess motivation questionnaires from the Amsterdam Chess Test (van der Maas & Wagenmakers, 2005). The verbal chess knowledge questionnaire characterizes the level of knowledge about four main aspects of the game, such as openings, positional and endgame principles, and visualization. The chess motivation questionnaire evaluates motivational aspects of the game. In addition, participants also completed the EPQ (Eysenck & Eysenck, 1984) and the emotional regulation questionnaire (Gross & John, 2003). The EPQ provides measures of the three main dimensions of the Eysenck PEN model: psychoticism, extraversion, and neuroticism. The emotional regulation questionnaire comprises two subscales: cognitive reappraisal and expressive suppression. The cognitive reappraisal subscale measures attempts to rebuild a situation that elicits a certain emotion, such that it modifies its impact in a significant way. The expressive suppression subscale is to do with the inhibition of expressing a given emotion. The first author of the study collected the data during the yearly chess competition of the Catalan Chess Federation. Each participant completed a self-report questionnaire comprising all tests. The time allocated for completing the verbal chess knowledge and chess motivation tests was controlled for each participant. All the participants signed a written informed consent disclaimer.

Table 7.2 compares the verbal chess knowledge, chess motivation, personality, emotional regulation, and Elo ratings of the studied sample with the respective measures in the wider population. These findings indicate that there were highly significant differences between the Dutch and Spanish chess players in the verbal chess knowledge test, albeit no meaningful differences concerning the chess motivation questionnaire. In addition, chess players scored significantly lower in the three scales from the EPQ: extraversion, neuroticism, and psychoticism. Regarding the emotional regulation strategies, while chess players scored similarly to the general population in cognitive reappraisal, they scored meaningfully higher than the general population did in the expressive suppression scale. There was minimal significance in the different Elo ratings between both groups of chess players (current study and ACT study), probably because the Elo ratings vary slightly across different chess federations. Effect sizes (*d*) were large for verbal chess knowledge and expressive suppression, medium for extraversion and neuroticism, and small for the rest of the variables (Cohen, 1988). These findings support the view that chess players are, on average, more introverted and more emotionally stable than the general population. Moreover, chess players are also more skilful at the deliberate concealment of emotions, corroborating the outcomes with Spanish child chess players and Austrian chess players regarding emotional regulation (Grabner et al., 2007; Hernández & Rodríguez, 2006).

Figure 7.2 shows a structural equation model with observed variables (Bollen, 1989) depicting the prediction of chess skill measured with the Elo rating, by the ACT verbal chess knowledge test and chess motivation questionnaire, the EPQ personality factors (extraversion, neuroticism, and

Table 7.2 *Comparison of verbal chess knowledge, chess motivation, personality factors, emotional regulation, and the Elo rating in chess players and in the general population (normative data for extraversion, neuroticism, and psychoticism: n = 527; normative data for verbal chess knowledge, chess motivation and Elo rating: n = 259; normative data for cognitive reappraisal and expressive suppression: n = 1,483; chess players: n = 100)*

| Factor | General population | | Chess players | | | |
|---|---|---|---|---|---|---|
| | M | Sd | M | Sd | *t* | *d* |
| Verbal chess knowledge[a] | 10.2 | 2.8 | 8.16 | 2.64 | 6.29*** | .75 |
| Chess motivation[a] | 98.30 | 11.90 | 98.41 | 12.07 | .08 | .01 |
| Extraversion | 12.74 | 4.13 | 10.76 | 4.34 | 4.36*** | .47 |
| Neuroticism | 12.20 | 5.42 | 9.12 | 5.57 | 5.19*** | .56 |
| Psychoticism | 6.09 | 3.58 | 4.98 | 3.18 | 2.89** | .33 |
| Cognitive reappraisal | 4.60 | 0.94 | 4.49 | 0.78 | 1.14 | .13 |
| Expressive suppression | 3.64 | 1.11 | 4.51 | 1.24 | 7.53*** | .74 |
| Elo rating[a] | 1865 | 284 | 1936 | 154 | 2.37* | .31 |

*Notes:**$p < 0.05$; ** < 0.01; *** < 0.001; [a] the mean and standard deviations are taken from van der Maas and Wagenmakers (2005).

psychoticism), and the emotional regulation strategies (cognitive reappraisal and expressive suppression). Double-headed arrows represent correlations. One-headed arrows represent causal effects, which are analogous to regression coefficients usually reported in regression analyses. One important advantage of this method over multiple regression analyses used in other studies is that it permits controlling for the correlations arising among the predictor variables (Bilalić et al., 2007b; Grabner et al., 2007; van der Maas & Wagenmakers, 2005). This model was specified and estimated with the maximum likelihood robust method and the R software (R Development Core Team, 2015; Rosseel, 2012). The model was estimated in two steps. In the first step, all exogenous variables (predictors) were specified as correlated. In the second step, the model was re-estimated by setting to zero the correlations that yielded non-significant values in the first step. The reported findings correspond to the second estimation step.

As shown in the model in Figure 7.2, the consistent predictors of chess skill were age, knowledge about openings and endgames, and expressive suppression, which explain altogether around 40% of the variability in the Elo rating ($R^2 = 0.41$). The model fit indices suggest that the specified set of relationships represented the observed data adequately (Hu & Bentler, 1999), with a non-significant chi-square value ($\chi^2[45] = 52.18$), and fair additional fit indices (CFI

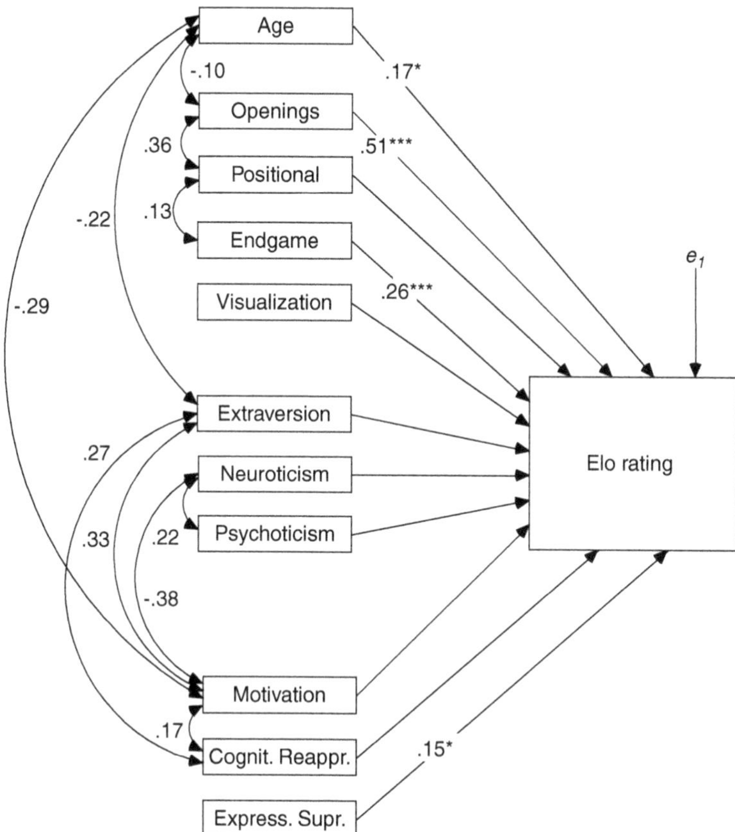


**Figure 7.2** Structural equation model with observed variables to predict the Elo rating, from age, verbal chess knowledge (openings, positional, endgame, visualization), personality factors (extraversion, neuroticism, psychoticism), chess motivation, and emotional regulation (cognitive reappraisal, expressive suppression); all correlation coefficients (double-headed arrows) were significant at the $p < 0.05$ level. The exogenous variable $e_1$ represents an error term

= 0.941, TLI = 0.914, and RMSEA = 0.040). Thus, these findings suggest that older, more experienced players with greater knowledge about openings and endgames, and who were more able to suppress emotions, were also those who had a higher level of chess skill as measured by the Elo chess rating. This is somewhat supportive of the interplay of emotional and cognitive systems for the activation of efficient chess thinking (Tikhomirov & Vinogradov, 1970). Moreover, these empirical findings are in accordance with those obtained with the Austrian chess players (Grabner et al., 2007), particularly concerning the effect of age, the lack of explanatory effects of personality variables analogous to those used in our study, and – more strikingly – the positive effects of control over the expression of emotions on chess skill. The Austrian players who reported a higher level of control over the expression of their own emotions also had a higher level of chess skill. Nevertheless, that study explains a slightly higher amount of the variability in the Elo chess rating (55%) compared with our own study, which may be because the study from Grabner and collaborators includes numerical intelligence measures that appear to be one of the stronger cognitive ability variables associated with chess skill (Burgoyne et al., 2016).

The Eysenck PEN model is a promising framework for addressing individual differences in personality related to individual differences in chess skill. The two main broad factors of this model, extraversion and neuroticism, have a robust biological basis and are directly comparable to other personality models. Even though neither factor contributed to explaining any variability in chess skill, extraversion and neuroticism were shown to be moderately higher in a sample of amateur chess players compared with the general population. The neuroticism factor relates closely to the emotional lability of the autonomic nervous system, whereas the extraversion factor relates to the degree of excitation and inhibition in the ARAS. Thus, introverted subjects might be characterized by stronger excitatory potentials of arousal or activation of the cerebral cortex and by weaker inhibitory potentials. The higher excitation prevailing in introverted subjects may therefore translate into greater resistance to fatigue, and a better ability to concentrate their attention. This may represent a strategic advantage for chess players in their field. Another advantage for chess players might stem from more straightforward management of emotional regulation. Expert acquisition and development in chess is a complex multifactorial phenomenon, however, addressed in the next chapter.

# 8

# Expertise

One of the first accounts attempting to describe the process of chess learning was provided at the beginning of the twentieth century (Cleveland, 1907). In this view, learning how to play chess involves a five-stage process of growing complexity and difficulty:

1. learning the names and movements of the pieces;
2. indefinite play, blundering, concentration on single moves or particular situations that do not lead to any systematic planning;
3. learning the strength and weaknesses of the pieces when acting in a coordinated way;
4. systematic planning of the development of the pieces; and
5. increasing power of analyses and 'positional sense'.

There are remarkable individual differences in the aforementioned scheme concerning the development and learning process of chess playing. People fortunate enough to reach an optimal performance in the fifth stage, can be considered as having achieved a chess expertise level. What are the characteristics that qualify a person as an expert, however? For example, one important characteristic that has recently been proposed is that an expert obtains results considerably above those obtained by the majority of people (Gobet, 2016).

Two main approaches are proposed for how to study expertise: absolute and relative. The absolute approach consists in analysing exceptional people in order to understand their endeavours and achievements in their respective domains. The relative approach consists in comparing the differences between experts and novices so as to understand what the determinants of expert performance are. Several methods are available to study expertise from both the absolute and the relative approaches. For example, an appropriate method to study expertise from the absolute approach is the historiometric method (Simonton, 2006). This method has been applied to the aesthetic success of classical composers (Simonton, 2000), the career landmarks of scientists and inventors (Simonton, 1997), and the educational and work achievements of extremely intelligent people (Lubinski, Webb, et al., 2001). From the relative approach, there are a number of experimental paradigms probably better suited to the analysis of expertise by comparing experts with novices (Chi, 2006b), such as the categorization of hierarchies of objects (Tanaka & Taylor, 1991). Furthermore, the structure and development of expertise can also be addressed with psychometric methods applied to study individual differences in chess expertise, from both the absolute (Blanch, 2018) and the relative approaches (Grabner et al., 2007).

In the light of these two approaches to expertise, the manifestation of expert behaviour emerges in both achievements and shortcomings, as shown in the top part of Table 8.1 (Chi, 2006a). Consider, for instance, the shortcoming of inflexibility of thought. Data from expert chess players indicate a positive association of expertise with thought flexibility, whereas inflexibility of thought was observed among ordinary experts when compared with higher-level experts. This phenomenon, termed the Einstellung effect, is used to explain the fact that more time was wasted on relevant chessboard squares in searching for familiar solutions than on the chessboard squares that were actually relevant for the most optimal solution (Bilalić, McLeod, & Gobet, 2008a; Bilalić et al., 2008b).

Furthermore, the performance of experts is likely to vary according to differences across domains. Good or poor expert performance depends on nine main characteristics of the domain, as shown in the bottom part of Table 8.1 (Shanteau, 1992, 2015). The improvement of individual competences within

Table 8.1 *Achievements and shortcomings of experts (Chi, 2006a), and good performance and poor performance of experts in accordance with task characteristics (Shanteau, 1992, 2015)*

| Achievements/good performance | Shortcomings/poor performance |
|---|---|
| Generation of best solutions | Expertise is limited to a domain |
| Detection of particular features | Overestimation of capabilities |
| Qualitative analysis of problems | Neglecting irrelevant details |
| Self-monitoring skills | Dependence on domain information |
| Selection of strategies | Inflexibility of thought |
| Opportunistic in using information | Wrong prediction of novice behaviour |
| Minimal cognitive effort | Bias and functional fixedness |
| Static stimuli | Dynamic stimuli |
| Decisions about things | Decisions about behaviour |
| Experts agree on stimuli | Experts disagree on stimuli |
| Predictable problems | Unpredictable problems |
| Tolerable errors | Intolerable errors |
| Repetitive tasks | Unique tasks |
| Availability of feedback | Unavailability of feedback |
| Decomposition of problems | Global problems |
| Use of decision aids | Lack of decision aids |

a domain depends on the characteristics of the respective domain. Some domains require lower levels of skill, however, because of the difficulties in measuring performance, poor feedback about performance, and lack of accurate assessment tools (Thomas & Lawrence, 2018). In contrast, chess experts have been deemed to be good performers (Shanteau, 1992), probably because performance is accurately measured with the Elo chess rating, and because of the immediate availability of feedback about performance after winning, drawing, or losing a chess game.

Chapter 5 highlighted that the brain structure and functioning of chess experts differ markedly from that of chess novices (see Table 5.1). These individual differences involve brain activation patterns, the integration of more brain areas useful for chess playing, the specialization of brain hemispheres, and the anatomical differences associated with the role of development and the starting age. Furthermore, these findings, together with the high stability and heritability of intelligence across the lifespan (Plomin & Petrill, 1997; Plomin & Spinath, 2004), suggest that chess expertise would be highly determined by innate factors, with minimal effects from practice. This point has been discussed extensively in the framework of the neural efficiency hypothesis regarding the effects of practice and expertise in chess (Grabner et al., 2006; Neubauer & Fink, 2009a). The level of expertise measured with the Elo rating is found to be positively correlated with speed, memory, and reasoning performance. More skilled chess players, however, showed higher cortical activation in the parietal lobes, though there was lower cortical activation in the frontal lobes in the speed and reasoning tasks. This finding supports the view that, aside from the acquired knowledge specific to the domain of chess, the general efficiency of the information-processing system is also important for cognitive performance. Hence, there are two fundamental questions to address in connection with individual differences in expert performance. The first question is whether practice increases or decreases the inter-individual variability in performance. The second question is to ascertain what the most influential factors are in terms of securing expert development.

As shown in Table 8.2, both questions have been argued to hinge on the type of task, such as whether it mainly requires either speed and accuracy skills or domain knowledge (Ackerman, 1987, 2007). For tasks implying speed or accuracy of motor movements, increments in practice lead to a reduction of individual differences in performance, with the main determinants of expert performance being perceptual speed and psychomotor abilities. In contrast, for tasks implying domain knowledge, increments in practice lead to an increment of individual differences in performance, with the main determinants of expert performance being general crystallized abilities, but also non-ability traits such as interests, personality, and motivation. The traits that determine expert performance in tasks involving domain knowledge, such as

Table 8.2 *Practice impact and determinants of individual differences in expert performance in accordance with the type of task (Ackerman, 2007). Differential effective strategies refer to tasks that can be completed successfully with different strategies. Inconsistent information processing refers to acquiring different skills to complete the same task. Closed tasks are bounded by a finite domain of knowledge, whereas open tasks are unbounded and cumulative*

| Type of task | Individual differences after the impact of practice | Determinants of individual differences |
|---|---|---|
| Speed/accuracy of motor movements | Reduced | Perceptual speed abilities<br>Psychomotor abilities |
| [Differential effective strategies, or inconsistent information processing] | Increased or remain constant | |
| Domain knowledge | Increased | General crystallized abilities (vocabulary, general knowledge, . . .) |
| [Closed or open] | | Interests<br>Personality<br>Motivation |

in the chess environment, are more varied through being more highly distributed in the brain than perceptual and motor skills. Moreover and apart from intellectual abilities, non-ability factors, together with an investment of time and effort in the domain, are also necessary for becoming an expert, as outlined by the PPIK theory (Ackerman, 2007, 2011). In any case, the role of practice is a central topic when attempting to explain the development of chess expertise.

## 8.1 The Role of Practice

As in many other fields, chess expertise can be attained only after a prolonged period of deep study and practice in the domain, which has been estimated to be around ten years (Ericsson & Lehman, 1996). A minimum of 3,000 hours with an average of 10,000 hours of practice have been judged as necessary to achieve expert performance in chess (Campitelli & Gobet, 2011; Simon & Chase, 1973). The gains in performance might be constrained, however, by what has been termed the power law of practice (Newell & Rosenbloom, 1981).

The simplest function that describes the power law is $T = a{\cdot}P^{-b}$, where $T$ is a performance measure describing the time invested in completing a task, $P$ is the practice measure (i.e., the trials in a task), and $a$ and $-b$ are empirical parameters indicating the performance time in the first trial ($a$) and the learning rate ($b$). One of the essential tenets of the power law of practice is that performance improves with increasing practice, which, beyond a given point, leads to progressively lower increments in performance. Figure 8.1 shows this behaviour in six tasks for 100 trials: mirror tracing, reading inverted text, scanning visual targets, sentence recognition, an online editing routine, and geometry proof justification. The parameters to build these curves were those provided in the study by Newell and Rosenbloom. As can be seen, the time invested in completing the task decreases progressively with the amount of trials invested in the task.

Even though these functions correspond to simple perceptual and motor tasks, the power law of practice could be extrapolated to other human behaviour. Nonetheless, learning complex academic concepts related to science, reading, and mathematics follows non-linear curves that may differ to a great extent from models following the power law (Cameron et al., 2015; Newell & Rosenbloom, 1981). Similarly, learning chess also involves a more complex set of intellectual skills, including the acquisition of extensive domain knowledge. Hence, there have been some attempts to explain chess skill acquisition by linking practice and performance with other mathematical functions aside from the power law of practice.

For example, learning curves following a negative exponential function have been suggested as having better explanatory power for early chess performance compared with learning curves following a power function (Gaschler et al., 2014). These findings are consistent with the idea that the early chess learning stages are more relevant for predicting individual differences in chess performance in the long term. A comparison of power, exponential, logarithmic, and quadratic functions with data from chess grandmasters suggests remarkable inter-individual variability (Howard, 2014c). Models relating practice and performance, such as those described by the power law of practice, might fail to predict performance in more complex activities. More specifically, the power function is found to be the best fitting curve for the most talented players, whereas a quadratic function is the best fitting curve for the less talented players. Additional findings from a group of novice players about the learning of chess endgames underline the fact that the judgement of learning does not improve meaningfully the learning performance (de Bruin, Rikers, & Schmidt, 2005), supporting the importance of prior knowledge for performance. Specific acquired domain knowledge ameliorates information-processing requirements to a great extent.

Practice tends to equate individual differences in performance when reaction time is the unit of performance. Nonetheless, the aforementioned analyses of learning curves with chess data highlight the fact that low- and high-ability subjects tend to generate substantially divergent learning curves, corroborating the association of skill learning with the type of task (Ackerman, 1987, 2007). In a complex intellectual task such as chess, practice activities should be

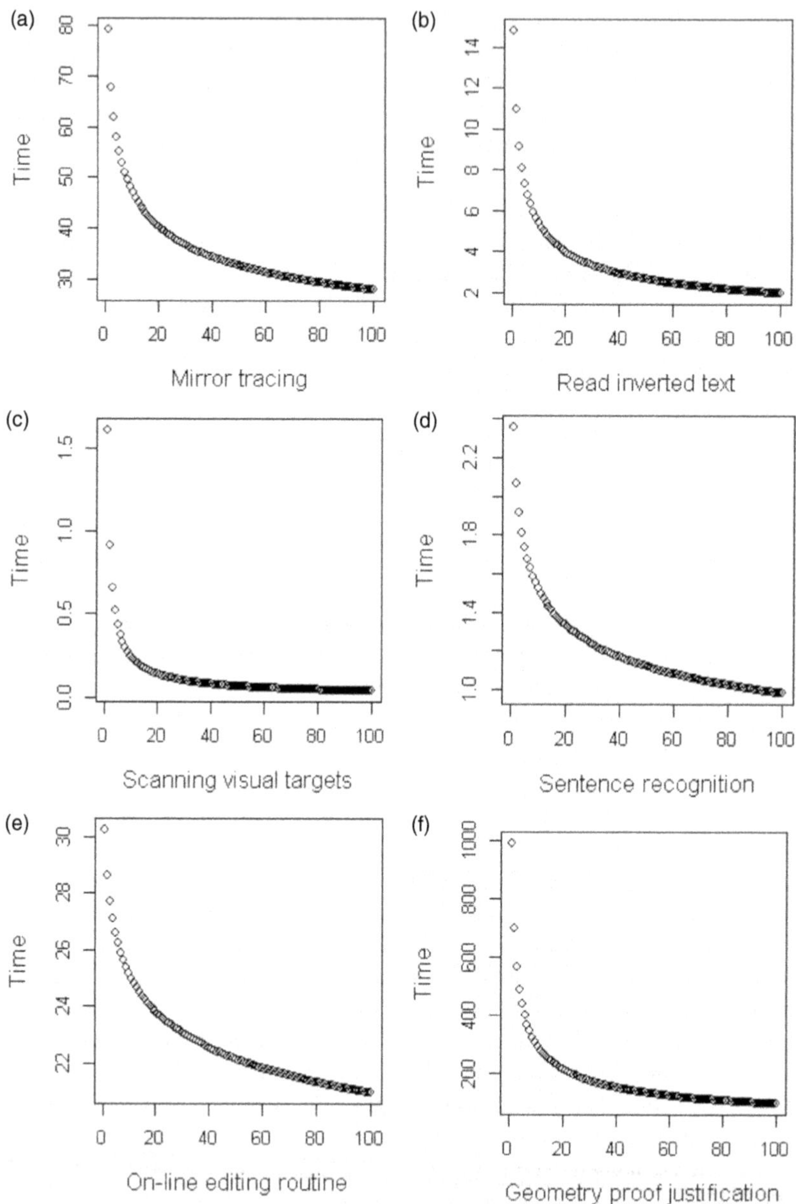


**Figure 8.1** The power law in action in six information-processing tasks: mirror tracing, reading inverted text, scanning visual targets, sentence recognition, an online editing routine, and geometry proof justification. The Y axis shows the time (seconds) invested in completing the task, the X axis represent 100 trials in each task. The empirical parameters (*a* and *b*) are those as suggested in the study by Newell and Rosenbloom (1981)

varied and complex as well. Moreover, becoming involved in practice activities might well depend on individual differences in intelligence, personality, motivation, or sex. More importantly, some crucial questions may be asked when considering practice in chess. What are these practice activities? Do they have equal influence in chess learning and in development towards chess expertise? Are there individual differences in the impact of practice activities on chess skill? The deliberate practice framework has addressed these questions extensively.

The deliberate practice framework maintains that the acquisition and development of expertise are, essentially, attributable to an assortment of activities over a prolonged period. These activities are reconciled with motivational and task requirements, and are, basically, designed to improve performance (Ericsson et al., 1993). Three main stages for the development of expertise have been proposed that encompass progressive increments in performance: the age at the initiation of practice; the transition to full-time involvement; and experts seeking eminent achievements. This three-stage process has been termed the monotonical relationship between accumulated deliberate practice and performance (Ericsson & Charness, 1994). Furthermore, the deliberate practice framework is explicitly reluctant to admit the influence of innate abilities for expert acquisition. On the other hand, it acknowledges that heritable individual differences in motivation and satisfaction with practice activities in the domain guide the degree of individual involvement in deliberate practice throughout the three aforementioned stages of expertise. The scrutiny of the deliberate practice framework in sport expertise has resulted in three central suggestions: the development of more nuanced theoretical assumptions, the improvement of research designs, and the modification of research questions (Baker & Young, 2014).

Deliberate practice has been suggested as a key factor for the acquisition of chess expertise, in accordance with empirical data. A study with two independent samples of chess players, from Canada, Germany, Russia, and the United States, explored whether chess practice – such as participating in chess tournaments, individual chess study, and coaching – were predictive of individual differences in chess skill (Charness et al., 2005). The findings from both samples corroborate the view that individual chess study is the most predictive variable of individual differences in chess skill, in addition to being more consistent beyond five years of serious play. In any case, only around 40% of the variability in chess skill is explained by chess practice activities. Another similar study, with a representative sample of eighty-one child chess players from the Netherlands, also looked at whether chess practice activities influence the development of chess skill over time (de Bruin et al., 2008). This study enquired into chess practice activities such as the number of chess coaches, books, CDs, and tournaments played per year, together with the total number of hours of serious play and study alone. The findings support the view that deliberate practice is positively associated with chess performance throughout the studied period. Besides, two groups of participants identified as either persistent or dropouts from the domain are found to benefit equally from deliberate practice, whereas gender has a smaller effect than deliberate practice in performance. Other factors are also likely to stimulate the degree of involvement in chess practice. For example, a high level of competitiveness and the will to excel are shown to exert strong influence on immersion in chess practice activities (de Bruin, Rikers, & Schmidt, 2007b). Empirical support for the deliberate practice framework has been far from unanimous or complete, however.

Guillermo Campitelli and Fernand Gobet address the association of practice activities with chess skill by using a longitudinal dataset of 104 chess players from Argentina (Campitelli & Gobet, 2008; Gobet & Campitelli, 2007). This research relates data about chess skill, measured by the national and international Elo rating, to individual and group practice, such as individual hours of study, individual and group coaching, and practising chess with other chess players, including games played in chess tournaments. In addition, the roles of the starting age and handedness were also examined. Higher involvement in individual and group practice activities is found to associate positively with chess skill. A substantial amount of the variability in chess skill is unexplained by the hours devoted to either individual or group practice, however. For example, some players needed only around 3,000 hours of effective practice to reach a master level, whereas other players needed over 23,000 hours to reach this same level (Gobet & Campitelli, 2007). Hence, other individual differences might account for the observed variability in chess skill. For instance, handedness and starting age relate in different degrees to chess skill. Chess players were more likely to be left handed than an equivalent sample from the general population, though the degree and direction of handedness were unrelated to chess skill. On the other hand, earlier involvement with chess practice activities was robustly related to chess skill. Therefore, the required skills to play chess well might be more easily acquired at earlier ages during a critical period, while increasing the amount of time available to build up a more extensive chess knowledge base.

Apart from the starting age, a relative age effect (RAE) or birthdate effect could in addition arise in the chess domain. The RAE entails the relationship of age to performance. For example, the youngest children in their year group at school and in sporting activities tend to achieve a lower level of performance than their classmates do. Children in the United Kingdom born in summer months show a disadvantage because the school year begins on 1 September, though the effect on the youngest children in the course group has also been observed in other countries where the academic course begins at different times of the year (Crawford, Dearden, & Meghir, 2007; Sykes, Bell, & Vidal Rodeiro, 2016). The population of expert chess players in the Northern

Hemisphere has been found to display a significant higher amount of births in the months of late winter and early spring – i.e., from January to April (Gobet & Chassy, 2008). A similar study corroborates these findings with several samples of chess players, additionally controlling for gender effects (Breznik & Law, 2016). These analyses included top junior males and females younger than twenty years old ($n = 367$ and $n = 387$, respectively), and top-level males and females ($n = 330$ and $n = 289$). Higher amounts of births in late winter and early spring were reported for all samples except for the top-level male sample.

These two studies from Campitelli and Gobet suggest little support for the extreme view of practice as the only fundamental requirement for reaching expert performance. The overall findings arguing against this deliberate practice extreme view led to the idea that deliberate practice is a necessary but not a sufficient condition to reach expert chess performance (Campitelli & Gobet, 2011). On the other hand, a limitation of the aforementioned studies reporting meaningful effects from practice on chess skill is that they omit measures of intelligence and personality. This may be an important reason as to why these predictive models explain only part of the variability in chess performance, with a substantial proportion of the variance being unexplained by practice-related factors only. For example, a more recent reanalysis of chess and music studies reports that, apart from deliberate practice, other candidate direct and indirect causal factors of performance could be the starting age, and also intelligence and personality (Hambrick et al., 2014). Because the deliberate practice framework has generally ignored the role of cognitive abilities in the structure and development of expertise, studies comparing the influence of both cognitive abilities and practice would help to elucidate the relative impact of the two domains on expert performance. In addressing the predictors of individual differences in chess skill, three of these studies consider the influence of individual differences not just on cognitive abilities but also on personality, motivation, and emotional regulation.

The most comprehensive predictive study is the research with the Austrian chess players, which includes several measures of individual differences in psychological variables other than deliberate practice (Grabner et al., 2007). The researchers evaluated the role of intelligence, personality, motivation, and emotional competences, and of chess attitudes and practice activities. The most predictive factors of individual differences in chess skill are numerical cognitive ability, age at entering the domain, current age, the number of rated tournament games played in the four years prior to the study, control over the expression of emotions, and motivation. All these factors explain 55% of the inter-individual variability in chess skill as measured by the Elo rating. Practice activities, such as practising alone with written material and computers, group practice, playing chess for fun, giving or receiving coaching, and watching tournaments in the media, are unrelated to chess skill. These null findings regarding the influence of deliberate practice on chess skill are partly attributed to the practice measures being unable to reflect actual engagement in deliberate practice. In contrast, the number of tournament games has a meaningful effect on chess skill.

Another study, undertaken with fifty-seven child chess players from the United Kingdom (Bilalić et al., 2007a), evaluates the influence of cognitive abilities, practice, chess experience, and age and gender on three measures of chess skill: a chess test, a recall task, and the knights row task. The main findings confirm that practice and experience in chess are better predictors of chess skill than intelligence, explaining between 69 and 86% of the variability in the three measures of skill. An elite chess subsample of twenty-three boys regularly participating in competitive chess had, on average, a significantly higher IQ than the children in the rest of the sample, even though IQ was negatively related with chess skill. In this subsample, the better players were those with lower IQs, with practice, experience, and age being – again – the stronger predictors of chess skill, and explaining 70% of the variability in chess skill. In any event, the study shows, again, that factors such as age, practice, experience, and intelligence might interact somehow in predicting chess skill, an idea advanced in the study by the same authors looking at the personality profiles of young chess players (Bilalić et al., 2007b).

The potential interaction of these sorts of variables was indeed addressed in another study with twenty-three child chess players from the Netherlands (de Bruin et al., 2014). The study evaluates intelligence (WISC), chess practice, chess enjoyment, and chess skill, as measured with an ad hoc chess test because the participants were beginners undergoing a chess course. Here, the more consistent predictors of chess skill are IQ and, to a lesser extent, the time devoted to chess practice activities outside the chess course, which explains 38% of the variability in chess skill. In the light of the novice level of the participants, cognitive ability might be distinctly important in the early stages of chess learning and expert development. Further, motivation is meaningfully linked with practice, suggesting that practice mediates the association between motivation and chess skill.

Another issue that has stimulated a certain controversy revolves around what is considered a deliberate practice activity, and what should be used as the measure of chess practice. For example, deliberate practice has been distinguished from actual practice. Deliberate practice in chess includes coaching, reading books, studying facets of the game such as openings and endgames, analysing games from top chess players, and playing against a machine. On the other hand, actual practice includes playing chess games against other humans (Howard, 2009). Deliberate practice estimates can be obtained by asking participants about their involvement in such practice activities with the retrospective recall method. Actual practice is usually estimated by computing the number of chess games within a given period. It has been argued that estimating deliberate practice can be difficult because of the inaccuracy at recalling career data (Howard, 2011b). In contrast, the number of games for different playing periods is usually kept within most chess rating lists. Some research works have compared whether deliberate practice or the number of games played is more predictive of chess skill. Data from the development of the Amsterdam Chess Test indicate that the number of official games played the year before a tournament is more predictive of the Elo rating than the estimated number of hours allotted to chess per week (van der Maas & Wagenmakers, 2005). In addition, the number of tournament games has been shown to be a stronger predictor of chess skill than practising with written material or computers, coaching, or watching games of other players, suggesting that deliberate practice estimates could be unreliably measured (Grabner et al., 2007). Moreover, the number of seasonal chess games has been related to sex differences in chess skill with data from chess tournaments (Blanch, Aluja, & Cornadó, 2015), or to the longitudinal change in the chess skill of top expert players (Blanch, 2018).

Further extensive work comprising four different studies with longitudinal data from several samples of chess players compares chess study hours, formal coaching, and the number of tournament games in the prediction of chess skill measured by the Elo rating (Howard, 2012a). The first study ($n = 533$) suggests that the number of games played is the strongest predictor of chess skill, followed by the number of study hours, and with the absence of effects of coaching. The second study compares participants with coaching ($n = 385$) and participants with no coaching ($n = 288$), indicating a longitudinal positive effect of coaching. The third study compares participants with high practice ($n = 39$) and participants with a low practice ($n = 38$), suggesting that the number of study hours is a weak predictor of chess skill with a low impact when controlling for the number of games and persistence in the domain. Finally, the fourth study compares five different samples of participants who had played a variable number of games, albeit with a constant persistence in the domain. This latter study supports the view that the number of games is a strong predictor of chess skill. Subsequent similar work by Robert Howard, including multiple studies and analysing several samples of chess players, reaches analogous conclusions regarding the number of games (Howard, 2013). In accordance with this set of studies, the number of rated chess games, which implies playing chess tournaments frequently, is the strongest causal factor of chess skill development over and above other forms of deliberate practice, such as chess study hours or coaching. This point of view has been criticized and disputed in greater depth, however, with a focus on the deliberate practice approach (Ericsson & Moxley, 2012; Howard, 2012b).

The deliberate practice approach has postulated the view that individual differences in expert performance are fully contingent on a variety of practice activities (Ericsson & Charness, 1994; Ericsson et al., 1993). Nevertheless, one of the crucial findings when examining the potential predictors of expert chess performance is that only a limited percentage of the variability in performance is due to practice-related factors. Hence, attributing the variability in expert chess performance to practice alone is highly misleading. Practice is obviously necessary to reach an expert chess level, but talent is also an essential component to account for the acquisition and development of expertise (Campitelli & Gobet, 2011). The talent versus practice debate is germane with the nature versus nurture debate addressed earlier, within Chapter 4. Chess has emerged as an excellent model with which to address the interplay of the multiplicity of intervening factors in the structure and development of expertise in other domains implying human intellectual activity, and, ultimately, to address the interplay of heredity and environment in this process (Gobet & Charness, 2006).

## 8.2 Talent versus Practice

Even though the nature versus nurture debate may be already over (Buss, 2001; Pinker, 2002; Tooby et al., 2003), whether talent or practice underlies expert performance has probably been one of the most contentious issues when studying individual differences in expert chess performance (Gobet, 2016). That at least 3,000 hours is the minimum amount of practice needed to accomplish an expert level might be true only for some individuals (Campitelli & Gobet, 2011; Simon & Chase, 1973). Some people are likely to become experts in the field with a lower amount of practising hours than these estimations within a somewhat shorter period. Other people may reach an expert level in chess after a relatively long period of considerable effort and sacrifice of other activities in life. Still other people will never reach an expert level no matter the time they invest in studying, coaching, or playing tournament games. A vast variety of gradients may fall in between, somewhere within these three main broad types. These different gradients reflect individual differences in expert performance, which, apart from practice activities, are governed by other individual differences in psychological traits and other factors.

In the seminal work about perception and chess skill from Herbert Simon and William Chase, practice in the form of thousands of hours is the answer to the question of how a person becomes an expert in chess (Simon & Chase, 1973). Nevertheless, these authors also acknowledge that practice should necessarily interact with talent, and that the combination of individual differences in certain cognitive abilities might be especially relevant for expert chess performance. The deliberate practice approach, in contrast, considers practice the key role to attain expert performance in a given domain. More specifically, individual differences in factors such as early experiences, preferences, opportunities, habits, training, and practice have been argued to be the crucial determinants of expert performance (Howe, Davidson, & Sloboda, 1998).

Deliberate practice comprises an assortment of activities undertaken on a voluntary basis to improve performance (Ericsson, 2006; Ericsson et al., 1993). The deliberate practice approach has taken at times an extreme environmental view, however, by excluding the influence of other individual differences on expert performance. On the other hand, there is an overwhelming amount of both conceptual and empirical evidence sharply opposing the deliberate practice approach.

A special issue in the journal *Intelligence* in 2014 ('Acquiring expertise: ability, practice, and other influences') addresses this topic in depth. To begin with, there are three main points to bear in mind with regard to expert performance, talent, and individual differences (Ackerman, 2014).

1. Practice is essential for expert performance.
2. Other factors explain the variability in expert performance apart from deliberate practice, such as physical limitations, injuries, early experiences, and ageing.
3. A substantial portion of the variability in expert performance is unexplained by practice among elite performers.

Another theme in the explicit review by Ackerman emphasizes that the deliberate practice approach tends to mistakenly identify talent with innate factors, which may be a highly spurious view. On the contrary, talent is viewed here as the result of the interaction between genes and environment, eventually leading to individual differences in the development of expertise (Ackerman, 2014). The same idea has also been commented on in depth in an empirical study addressing the issue of nature, nurture, and expertise (Plomin et al., 2014). From this point of view, although the development of expertise from the deliberate practice approach represents a rigid and passive acquisition process, the genotype–environment correlation approach renders a flexible and active model. Individuals select and create their own environments in accordance with their genetic endowment (Scarr & McCartney, 1983). The inflexibility of the deliberate practice approach, as advocated by Ericsson himself, becomes particularly apparent in the counter-argument to the effect that no specific theory or approach is special (Wai, 2014b). With regard to the development of expert performance in the domain of chess, it has been argued that Anders Ericsson interpreted the findings as being consistent with the deliberate practice approach, though he rejected the hypothesis of individual differences in cognitive abilities as explanatory factors of further variance in expert chess performance (Grabner, 2014b, 2014a). More specifically, it is argued that, in the analyses of chess expertise from the deliberate practice approach, there was a remarkably biased picture of empirical findings, implausible explanations of the main research findings, and neglect of part of the available evidence (Grabner, 2014a).

There are, in addition, consistent empirical findings running counter to the deliberate practice position when advocating practice as the sole causal factor of expert performance in several fields, including, of course, chess. This evidence has come from studies applying diverse research methods. For example, the combination of prospective and retrospective longitudinal data highlights the importance of cognitive abilities for the development of expertise in education and occupational domains (Wai, 2014a). The prospective data were from the Study of Mathematically Precocious Youth (SMPY, *n* = 1,975) and the Project Talent (*n* = 1,536). The retrospective data corresponded to five groups of the US elite (*n* = 2,254): Fortune 500 CEOs, federal judges, billionaires, senators, and members of the House of Representatives. The main findings are surprisingly similar across both educational and occupational samples, and support the view that, regardless of deliberate practice, cognitive ability – especially mathematical ability – plays a central role in succeeding in these domains.

Additional data obtained with a mixed method approach yielded similar findings with the third cohort from the SMPY study (Kell, Lubinski, & Benbow, 2013). A group of individuals with remarkable achievements in occupational fields such as business, law, medicine, and information technology had already been identified at a precocious age, below thirteen years, as being exceptional achievers in mathematical and verbal reasoning abilities. Another multivariate behavioural genetic study (Plomin et al., 2014) examined reading achievement with an extensive group of twin pairs (*n* = 4,955). Although a substantial part of the variability in reading performance was due to genetic factors (62%), only a modest part was due to common environmental influences shared by the participants (14%). These findings support the view that genetic factors indeed play a more crucial role than shared environmental influences on reading performance, and contradict to a great extent the deliberate practice approach.

Similar findings have been repeatedly reported with data from the chess domain in a diversity of empirical studies. For example, findings about the development of chess expertise in children, through several experiments and a respectable sample size, suggest that in no way might experience be able to substitute for talent (Horgan & Morgan, 1990), and that a high amount of practice would be unlikely to lead an ordinary player into becoming a chess master. Additional recent studies with children contrasting the effects of cognitive abilities and practice suggest that cognitive abilities are undoubtedly important for the development of chess expertise (Bilalić et al., 2007a; de Bruin et al., 2014), findings corroborated with data from adult chess players (Grabner et al., 2007).

Detailed comprehensive reports contrasting practice with other factors certainly suggest that practice is indeed a necessary but not a sufficient condition for achieving expert chess performance (Campitelli & Gobet, 2011), a point that has been additionally supported when analysing elite chess experts. In general, this body of evidence has highlighted the importance of factors other than practice for expert performance. These studies have analysed elite chess players in the expertise framework, aiming to compare the relative contribution of practice in contrast to other factors. The findings from elite chess players are particularly relevant because they are based on significant samples composed of individuals with outstanding achievements (Simonton, 1999, 2006). For example, a notable study examined the case of the Polgar sisters, three female chess players who reached an outstanding chess level (Howard, 2011a). This study asked specifically whether intellectual performance in chess depended only on practice. The three Polgar sisters learned chess at a very young age as part of their home school programme, being exposed to a considerable amount of daily chess practice and deep chess analysis. With equivalent levels of practice, were that all that mattered for expert performance the three sisters should have reached similar levels of expertise along their chess careers; but this was not the case. Only one sister reached the top level of expert performance, becoming the only woman in chess history to have been ranked within the top ten players in the world. This was taken as supportive evidence for the role of natural talent for expert chess performance.

Another study contradicting the deliberate practice approach examined in detail the case of Magnus Carlsen, the current world chess champion occupying the number one spot in the December 2018 FIDE rating list, with 2835 Elo points (Gobet & Ereku, 2014). Data about the Elo rating, starting age, current age, and number of years of practice were compared between the eleven top players in the January 2014 FIDE ranking list. There was a statistically significant difference between the ratings from Magnus Carlsen and the mean rating of the other ten players, and also in the estimated number of years of practice. Because Carlsen had the highest Elo rating even though he had fewer years of practice than his colleagues, deliberate practice is argued to be unable to explain these findings in accordance with its main tenets. Natural talent was deemed to be another plausible and more likely explanation for these findings. A more recent study conducted a longitudinal examination of the top 100 chess experts in the world, from 2009 to 2015 (Blanch, 2018). This study undertook a cross-domain analysis of change over time, examining the extent to which age and the change in the number of games could explain the change in the Elo ratings throughout the observed period. These findings show that half the variation in the change in chess performance during the period is unexplained by the combination of age and the change in tournament activity. Playing more games did not contribute to additional gains in chess skill, contradicting the extreme deliberate practice view.

These latter findings are consistent with meta-analytical research examining the explained variability in performance by deliberate practice. Recent meta-analyses from several domains have suggested that expert performance is poorly explained by deliberate practice alone. Two separate meta-analyses examined deliberate practice in sports (Macnamara, Moreau, & Hambrick, 2016) and in music, games, sports, education, and the professions (Macnamara, Hambrick, & Oswald, 2014). The findings from both the studies oppose the claims raised by the deliberate practice approach, with very large amounts of the variability in performance being unexplained by practice. Just 18% of the variability in overall sport performance is explained by practice, and, more importantly, only a meagre 1% when considering elite samples. Similarly, the unexplained variance in performance by deliberate practice in music, games, sports, education, and the professions is much larger than the explained variance, which ranges from only 26% for games to less than 1% for the professions. Another meta-analysis focused in particular on two of the domains that have generated larger bodies of research into expertise: chess and music (Hambrick et al., 2014). The figures here are a little more optimistic regarding the predictive ability of deliberate practice, even though they are far from the extreme position, which attributes the full variability in performance to practice. The explained variability in performance by deliberate practice was found to be 34% for chess and 30% for music. A more recent meta-analysis about the academic selection hypothesis also seems to provide support for the role of cognitive abilities in the acquisition of expert performance in chess (Sala, Burgoyne, et al., 2017). Indeed, a remarkable and surprising fact derived from this body of knowledge is that deliberate practice is considered to be important for expert chess performance, unlike the radical version of the deliberate practice approach, which denies the influence of other, well-established factors. Altogether, the body of research reviewed so far within this section emphasizes that, although deliberate practice plays a role in the development of expertise, there are several other relevant factors, largely supported by consistent scientific evidence. What are these factors, and how do they interrelate with deliberate practice?

Comprehensive models about the development of expertise take into account the integration of multiple genetic and environmental factors, including deliberate practice, and represent a more reasonable and straightforward approach to explaining expert performance, in chess and in other fields. For example, the research agenda proposed by Dean Simonton (Simonton, 2014a) comprises a series of direct and indirect causal links between six broad components: genetic factors, environmental factors, cognitive abilities, dispositional traits, deliberate practice, and creative performance. Genetic and environmental factors act as distal causes of creative performance, and as direct causes of both cognitive abilities and dispositional traits. Cognitive abilities and dispositional traits are, in turn, linked with deliberate practice, and with creative performance. Deliberate practice is also linked with creative performance. Furthermore, Simonton has advanced four central questions to guide empirical studies (Simonton, 2014b). These four questions are as follows.

1. Is creative performance fully explained by deliberate practice?
2. Is creative performance still fully explained by deliberate practice after considering cognitive abilities and dispositional traits?
3. Are there other predictors of the cross-sectional variability in creative performance?
4. Do the variables in the third question underlie genetic factors?

In this view, negative answers to questions (1) and (2), or affirmative answers to questions (3) and (4), would seriously challenge the deliberate practice framework. Furthermore, the multifactorial gene–environment interaction model (MGIM) involves very similar elements to those suggested in Simonton's agenda (Ullén et al., 2016). Genetic and environmental factors influence phenotype-level individual differences, such as abilities, personality, interests, and motivation. These phenotypic variables associate with neural mechanisms and deliberate practice, the latter influencing in turn neural mechanisms and physical properties, which eventually may lead to the development of expertise. Simonton's research agenda and the MGIM are readily extrapolatable to the chess domain to address the role of deliberate practice in expert chess performance. In fact, both conceptual models support several of the predictions and the empirical findings reviewed so far regarding chess and individual differences.

There are two additional central issues in connection with expert performance: ageing and sex differences. Both have been stressed in the programme of research developed by Ackerman (Ackerman, 2007), and they are indeed relevant in the chess domain. First, middle-aged chess players generally have a more extensive knowledge base than young chess players have. Second, there are remarkable gender differences regarding participation and performance in the chess domains. The first of these points, dealing with age differences, is addressed in the next section. The second point, dealing with sex differences, is also addressed in the next section with a comparison involving age effects, and in greater depth in the next chapter.

## 8.3 Cognitive Decline in Chess

Ageing is a crucial factor in a cognitively demanding domain such as chess. Even though several cognitive abilities may remain fairly stable from infancy to old age (Deary et al., 2000), the key cognitive abilities required for chess also tend to decline through the lifespan. There are notable individual differences in this process, however. The fluid and crystallized model of intelligence (*Gf* ~ *Gc*) is appealing in order to explain crucial changes in these individual differences in cognitive abilities and performance (Baltes, 1987; Cattell, 1963, 1987).

In this approach, abilities related to the biological basis of intelligence (*Gf*) are viewed as declining earlier and more sharply than abilities related to cultural contents and acquired knowledge and skills (*Gc*), which tend to remain more stable, and even grow over time under certain circumstances. One of the most comprehensive studies addressing adult intellectual development followed up more than 5,000 individuals for thirty-five years, from 1956 to 1991 (Schaie, 1994). This study points out several protective factors in connection with cognitive decline, such as a healthy lifestyle and life satisfaction, living in a favourable environment, a flexible personality, being married to somebody with a high cognitive level, maintaining higher levels of perceptual processing, and involvement in stimulating intellectual activities. Aside from some of these characteristics, chess is intellectually demanding and cognitively stimulating. Thus, several studies have enquired into the magnitude of the cognitive decline in chess experts.

In the past few decades increasingly young players have been ruling the chess domain (see Table 6.3). Since 1795 younger and younger players have been winning the World Chess Championship, with a negative correlation between the year of achieving the championship and the respective champion's age. Young and very young chess players are nowadays achieving a remarkable level of chess performance, particularly at top elite chess (Gobet et al., 2002; Howard, 1999, 2005b), an effect likely to depend on natural talent rather than on changes in the chess environment. Nonetheless, people at old ages can play chess very well, and retiring from this domain generally occurs at meaningfully later ages than in other sports. For example, there is a much slower decline in chess performance compared with physical activities such as running or swimming (Fair, 2007). A model involving the exponential growth and exponential decline of lifetime performance suggests that the age of peak performance is twenty-one years for swimming, twenty-six years for track and field, and thirty-one years for chess (Berthelot et al., 2012). An analysis of world chess champions by age at the time of the critical world championship match from 1858 to 1960 highlights that ageing beyond forty years old might, however, decrease the quality and performance in competitive chess (Rubin, 1960). On the other hand, another statistical analysis of world chess champions, together with additional personal career data, suggests that the mental abilities of professional chess players are unhampered by ageing (Draper, 1963). Nevertheless, older expert chess players might experience more difficulties in defeating younger chess players because of their playing peculiarities becoming increasingly known, and because of a higher susceptibility to fatigue affecting crucial chess-playing decisions, such as accepting a higher number of draws. Further accounts about the influence of ageing on thirty-two famous and strong chess players from all times suggest, for instance, that Akiba Rubinstein's (1882–1961) decline began at forty-three years of age, and that Savielly Tartakover's (1887–1956) decline began at fifty-nine years (Krogius,

1976). Krogius also underlines the importance of an early start in chess, arguing that chess players starting at earlier ages tend to have a longer active playing career, a claim that is substantiated in a subsequent empirical study focused on the role of practice in the development of chess skill (Gobet & Campitelli, 2007).

In fact, age is associated with the level of chess expertise and other factors in intriguing ways. For example, it has been found that child chess experts have superior performance to adult chess novices in the immediate recall of chess positions, yet inferior performance in a digit span memory test (Chi, 1978; Schneider et al., 1993). Older players, however, appear to match the problem-solving abilities of younger players with equivalent levels of chess skill, even with constraints in information encoding and retrieval. Moreover, older players tend to invest a smaller amount of time than younger players in selecting optimal moves, probably through being able to apply more refined move generation processes (Charness, 1981). A meta-analysis with four studies finds both skill and age effects in two types of common experimental tasks in chess research: the best move task and the recall task (Moxley & Charness, 2013). Age associated negatively with performance in the two tasks, whereas skill associated positively with performance in the two tasks. Thus, crystallized knowledge might be particularly influential concerning the memorization of chess positions. In addition, a comparison involving two groups of young ($n$ = 29) and old chess players ($n$ = 30) reports that chess players at higher levels of expertise responded more quickly in detecting chess threats, even though the level of chess skill did not moderate the age effects on the speed of information processing (Jastrzembski et al., 2006). Another study with a group of forty-four tournament-level chess players examined verbalized operational senses (VOSs) – i.e., individual differences in thinking about an object (Vasyukova, 2012). VOSs consist of several indicators derived from thinking aloud verbal reports from individuals when aiming to select the best move for a given chess position. Players below forty years of age show a deeper transfer of VOSs than players over forty.

Age, ability, and productivity are indeed largely intertwined. A large study with over 40,000 chess players analysed FIDE worldwide tournaments between 2008 and 2011, examining the association of age with mental productivity (Bertoni, Brunello, & Rocco, 2015). Less talented players were more prone to drop out with ageing and also bore a more pronounced selection effect at earlier ages. Besides, older chess experts (forty years old and above) displayed a 10% decrease in productivity, while productivity was remarkably high for younger players (fifteen years of age). Interestingly, this econometric study interprets the findings in terms of the relation between fluid and crystallized intelligence. Crystallized intelligence was unable to compensate for the early decline in fluid intelligence, which supported somehow the importance of fluid abilities for chess performance. In contrast, computational findings derived from a neural network approach suggest that pre-existing crystallized knowledge might compensate well for the age decline on memory span performance (Mireles & Charness, 2002).

Taken together, these latter findings support the view that an extensive knowledge base is one of the most important elements for chess expertise. Such a knowledge base can be acquired only after several years of intensive practice and study. Two large-scale research works have addressed the role of this knowledge base considering the crucial factor of tournament activity and its relationship to ageing. Both studies address this topic by evaluating the hypothesis that age is kinder to the more able. This hypothesis maintains that the age decline in performance tends to be less marked in those people with higher intellectual performance than in those people with lower intellectual performance. The first study was undertaken with a sample of strong chess players ($n$ = 5,011) with a mean Elo rating of 2173 points ($Sd$ = 117) and a multi-level modelling technique (Roring & Charness, 2007). This study reports a later peak performance independent of the initial level of chess skill, supporting the hypothesis that age is kinder to the more able. The decline in performance was moderate beyond their peak for those players with a higher initial performance than for those players with a lower initial performance. Furthermore, tournament activity related to performance while interacting with age, albeit with smaller effects on performance for the older players.

The second study undertook a comprehensive analysis of the hypothesis that age is kinder to the more able with two sizeable samples ($n$ = 100,529 and $n$ = 119,785) from the FIDE and the German Chess Federation (DSB), respectively (Vaci, Gula, & Bilalić, 2015). This assortment of data was modelled in accordance with Simonton's model of career trajectories and landmarks (Simonton, 1997). This model describes three different stages of the creative process during the individual's lifespan: (1) fast production of ideas; (2) productivity levelling off; and (3) post-peak decline in producing new ideas. Chess players reached their peak around thirty-eight years of age, with more able players starting to decline earlier (fifty-two years) than less able players (fifty-seven years). Nevertheless, tournament activity influenced these declines and the subsequent stabilization, because higher levels of tournament activity contributed to the holding of higher levels of skill at advanced ages.

Sex is another central factor that might intertwine with ageing when related to chess expertise. A large study with US players ($n$ = 256,741) suggests that boys began to compete in serious chess earlier than girls (Chabris & Glickman, 2006). Another study with data from chess tournaments ($n$ = 553, 42 females) reports a curvilinear U-shaped relationship of age to sex differences in the Elo ratings of males and females (Blanch et al., 2015). These findings suggest that sex differences in the Elo ratings are higher for both the youngest and the oldest females, who were also those less involved in chess tournament activity.

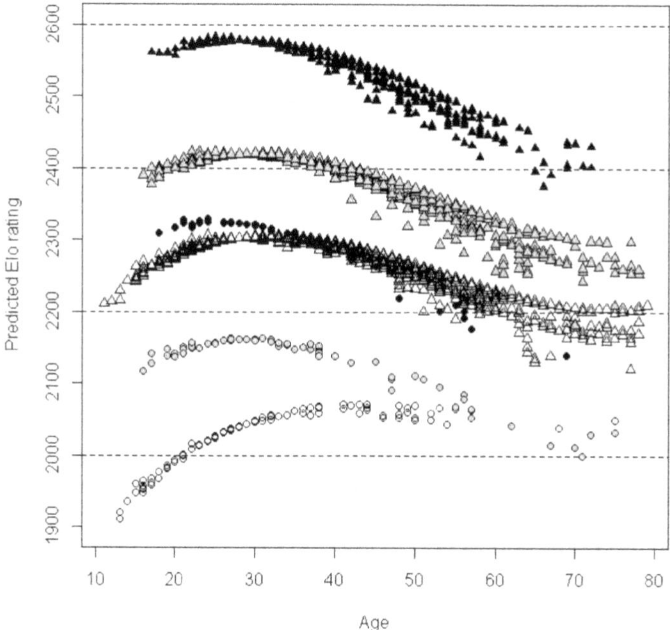


**Figure 8.2** Association of age with the predicted Elo rating at three levels of expertise: FIDE masters (white colouring), international masters (grey), and grandmasters (black). Triangles represent men, and circles represent women (data: FIDE list, March 2014)

Looking at the cross-sectional association of age, tournament activity, and their interaction with chess performance at different levels of expertise is revealing when analysing the data contained in the FIDE list of March 2014. There is nothing especially attractive about this particular list; it was just taken as an example. After selecting the titled active players between ten and eighty years of age, there were three main groups of experts: FIDE masters ($n$ = 1,379, 116 females), international masters ($n$ = 864, 88 females); and grandmasters ($n$ = 522, 47 females). Within the three groups, men had a higher average Elo rating and were older than women.

These data were additionally modelled with a regression analysis. The Elo rating was entered into the regression model as the dependent variable, while age, sex, chess title (FIDE, international master, or grandmaster), and number of games in the period, together with all possible interactions, were entered as the independent variables. In addition, age was modelled with a third polynomial (cubic function) in accordance with Simonton's model applied in the aforementioned study with German chess players (Simonton, 1997; Vaci et al., 2015). The regression model findings indicate a robust association of the independent variables and their interactions with the Elo rating. Approximately 75% of the variation in the Elo rating was explained by the combination of the independent variables ($R^2$ = 0.74). Most notably, when examining the association of age with the Elo rating at the three expert levels (FIDE, international master, and grandmaster), there is considerable variability regarding expertise and sex (see Figure 8.2). As expected, the predicted Elo rating was the lowest for the FIDE masters and the highest for the grandmasters, while also being generally lower beyond thirty or forty years of age. These findings suggest that the cognitive decline in chess arises independently of the expert level. What is more remarkable in these data, however, is that the female FIDE masters and the female international masters showed lower Elo ratings throughout the whole age span, whereas the curve for the female grandmasters overlapped largely with the curve of the male FIDE masters. This finding highlights that, on average, men exhibit higher chess performance across different skill and age levels than women. This outcome could be attributed to the overrepresentation of men compared to women in the chess domain.

What are the factors contributing to the underrepresentation of women in chess compared with men? Does this underrepresentation explain the comparatively low performance of women? Do men play like women? These questions stemming from the sharp disparity between men and women regarding the participation rates and performance in the chess domain constitute the topic addressed in the next chapter.

# 9

# Sex Differences

The evolution and expression of sex differences in psychological traits has been attributed to sexual selection and sex hormones (Geary, 2006, 2010). Sex hormones organize the neurocognitive structure of human beings at early stages of development, disputing the view of social forces as the sole mechanism shaping sex differences in psychological traits (Halpern, 1997; Halpern & Wright, 1996; Kimura, 1993, 1996). Sexual selection predicts sex differences in brain size, organization, and functioning, with empirical findings highlighting salient brain differences. Males have, on average, larger volumes in brain parameters, including grey and white matter (Ruigrok et al., 2014). Females have more white matter and less grey matter compared to men in areas related to general intelligence, as measured with the WAIS (Haier et al., 2005). IQ and grey matter correlations are strongest in frontal and parietal lobes for men, whereas, in contrast, IQ and white matter correlations are strongest in the frontal lobe and Broca's area for women. With equivalent IQ scores, both sexes achieve similar IQs with divergent brain designs.

Whether there are sex differences in psychological traits, however, has been argued to be of little scientific interest. In contrast, the origins, magnitude, and practical implications of these differences appear as more pressing issues within this field of research (Geary, 1995a, 2010). Substantial and practical reasons for studying sex differences in major psychological traits had already been advanced by pioneering work in differential psychology (Anastasi, 1937). At the time of the work from Anne Anastasi, sex differences favouring men in intellectual achievement were reported as being outrageously high, as a result of women experiencing considerable discrimination in several occupations and educational opportunities on the basis of gender alone. This seminal work raised several additional complications to studying sex differences in psychological traits, such as sampling and measurement issues, or the earlier maturation of girls over boys. Moreover, the higher variability of men over women and consideration of cultural factors are often intermingled in contemporary research.

There is allegedly higher variability for men compared to women regarding cognitive abilities, and to a lesser extent regarding personality factors. For example, the cross-cultural variability of men and women has been reported to be highly homogeneous in terms of verbal ability, albeit differing substantially in mathematical and spatial abilities (Feingold, 1994b). Further analyses with outstanding individuals from six national samples do indeed suggest greater variance for males in several cognitive abilities, excluding reading comprehension, perceptual speed, and associative memory (Hedges & Nowell, 1995). Additional evidence from two large-scale samples from Scotland also points to higher variability for men than women in general intelligence (Johnson, Carothers, & Deary, 2008). Men were found to be more variable than women at the lower-left tail, however, though less variable at the upper-right tail of the score distribution. Higher cross-cultural variability for men compared to women has also been reported in the big five personality factors in a large cross-cultural study ($n$ = 12,156). Higher proportions of men emerged in both tails of the distributions in several facets of the NEO-PI-R instrument (Borkenau, McCrae, & Terracciano, 2013).

Examination of cultural factors implies that if a psychological trait is genetically determined then sex differences in that trait should be invariable across different cultures. On the other hand, if a psychological trait is environmentally determined then sex differences in that trait should vary between different cultures. The development of sex differences in the big five personality traits has been found to be quite similar across twenty-three different cultures, supporting the arguments for a stronger hereditary basis for these traits (De Bolle et al., 2015). Indeed, the division and specialization of many activities between the sexes has been observed across different cultures while exhibiting a relative degree of universality. Many similar functions are performed by either men or women across different cultures (Buss, 2001). Nevertheless, a cross-cultural analysis of the behaviour of men and women documents profound changes in the psychological attributes and behaviour of women when assuming traditionally male-dominated roles revolving around five main themes (Wood & Eagly, 2002):

1. intensification of the value placed on job attributes such as freedom, challenge, leadership, prestige and power;
2. more balanced career aspirations, akin to those of men;
3. engagement in higher-risk behaviours;
4. increments in self-reports of assertiveness, dominance, and masculinity; and
5. higher equalization of scores in mathematics and science with those of men.

Sex differences in psychological traits continue to attract considerable attention from behavioural researchers from diverse disciplines (Geary, 2010; Gobet, 2016). More specifically, intelligence and personality, frequently associated with performance in academic subjects and occupational choices, are the two broad constructs that have elicited a respectable amount of empirical and theoretical studies in connection with expertise.

## 9.1 Sex Differences in Intelligence and Personality

The study of sex differences in cognitive abilities and personality dispositions is rife with challenging controversies. A comprehensive synthesis of the average differences commonly found between men and women summarizes a variety of cognitive tests and tasks usually showing significant sex differences (Halpern, 1997). Several implications for education have been derived from these findings. For instance, sex differences in cognitive abilities should not be used to restrict individual educational choices throughout development, and should not be interpreted in terms of better versus worse (see Table 9.1). For example, early findings from the Study of Mathematically Precocious Youth (SMPY) reported remarkable superiority for males over females in mathematical reasoning ability. This gap was attributed to a combination of hereditary and environmental variables, rather than being entirely attributable to socialization and educational processes (Benbow & Stanley, 1980).

More than two decades later, however, another report based on a comprehensive review of forty-six meta-analyses maintained that men and women are very similar on most psychological variables, a conjecture that has been termed the gender similarities hypothesis (Hyde, 2005). These meta-analyses were grouped into the evaluation of six main categories, namely cognitive variables (i.e., abilities), non-verbal communication, social or personality variables (e.g., aggression, leadership), psychological well-being (i.e., self-esteem), motor behaviour (e.g., throwing distance), and miscellaneous constructs (i.e., moral reasoning). The findings in this study give greater support to the gender similarities hypothesis, except for motor behaviours, some aspects of sexuality, and aggression.

Nonetheless, the evidence pointing to sex differences in cognitive abilities is quite extensive. Two meta-analyses have looked at whether there are sex differences in general intelligence (*g*). One meta-analysis comprised studies from the general population (Lynn & Irwing, 2004), whereas the other comprised studies with university students (Irwing & Lynn, 2005). The findings diverge from the sex similarities hypothesis concerning general intelligence, however, reporting a male advantage from fifteen years of age, together with a male advantage of five IQ points in mean scores, and inconclusive findings regarding the higher variability of men over women for the university students. Furthermore, and in line with the anatomical brain differences between men and women alluded to earlier (Haier et al., 2005; Ruigrok et al., 2014), sex differences are also reported when studying the neural efficiency hypothesis. Men and women with higher scores in cognitive abilities show higher brain efficiency when solving figural-spatial tasks, even though men have a meaningfully lower brain activation in frontal brain areas (Neubauer & Fink, 2009b). A further, more detailed review reports in addition that the NEH is more robustly supported for men than for women. This inconsistency is attributed to a larger decoupling of frontal brain areas for highly intelligent males, and a larger coupling between the frontal and parietal/occipital brain areas for highly intelligent women (Neubauer & Fink, 2009a). Sex differences in the NEH also emerge for different kinds of cognitive tasks, for men when performing a figural-spatial task, and for women when performing a verbal matching task. These findings corroborate the average sex differences in specific cognitive abilities summarized in Table 9.1.

Table 9.1 *Cognitive tasks and tests showing sex differences (adapted from Halpern, 1997)*

| Task/test | Examples |
|---|---|
| <u>Women obtain higher average scores</u> | |
| Rapid access to and use of phonological, semantic, and other information in long-term memory | Verbal fluency/phonological retrieval; synonym generation/meaning retrieval; associative memory; memory battery/multiple tests; spelling and anagrams; mathematical calculations; memory for spatial location; memory for odours |
| Knowledge areas | Literature; foreign languages |
| Production and comprehension of complex prose | Reading comprehension; writing |
| Fine motor tasks | Mirror tracing – novel, complex figures; pegboard tasks; matching and coding tasks |
| Perceptual speed | Multiple speeded tasks; 'finding "A"s' – an embedded letters test |
| Decoding non-verbal communication | – |
| Perceptual thresholds (large, varied literature with multiple modalities) | Touch – lower thresholds; taste – lower thresholds; hearing (males have greater hearing loss with age); odour – lower thresholds |
| Higher grades in school (all or most subjects) | – |
| Speech articulation | Tongue-twisters |
| <u>Men obtain higher average scores</u> | |
| Tasks that require transformations in visual working memory | Mental rotation; Piaget water level test |
| Tasks that involve moving objects | Dynamic spatio-temporal tasks |

Table 9.1 *Cont.*

| Task/test | Examples |
|---|---|
| Motor tasks that involve aiming | Accuracy in throwing balls or darts |
| Knowledge areas | General knowledge; geography; maths and science |
| Tests of fluid reasoning (especially in mathematical and science domains) | Proportional reasoning tasks; Scholastic Assessment Test/Mathematics Graduate Record Examination – quantitative; mechanical reasoning; verbal analogies; scientific reasoning |

Robust findings emerge in addition when examining sex differences in broad personality factors. One of the earliest meta-analyses examined a comprehensive set of studies from 1940 to 1992 (Feingold, 1994a). Males were found to be more assertive and higher in self-esteem than females, whereas females were found to be more extraverted, anxious, trustful, and tender-minded. Moreover, these sex differences hold across ages, educational level, and nations. More recent research works suggest that sex differences emerge during adolescence in the big five personality factors, with very similar patterns across different cultures. Girls score higher than boys do in neuroticism, which stabilizes by fourteen years of age. Girls also score higher than boys do in openness to experience and conscientiousness between twelve and seventeen years old. Further, sex differences in extraversion and agreeableness at seventeen years old tend to be already similar to those observed in adulthood. More interestingly, these sex differences in personality are strikingly analogous across cultures (De Bolle et al., 2015). Another large-scale study ($n$ = 55 cultures, $n$ = 17,637 participants) reports that women tend to score higher than men in neuroticism, extraversion, agreeableness, and conscientiousness (Schmitt et al., 2008). Moreover, this study argues that these sex differences in personality are significantly higher in egalitarian cultures and lower in more disadvantaged cultures.

In any event, a more recent study from the United States shows that women remained underrepresented in elite professions that are likely to require a high level of cognitive ability, such as federal judgeship, the House, the Senate, and – to a larger extent – billionaires and Fortune 500 CEOs. Whether the lower representation of women is actually due to differences in cognitive ability, however, is argued to be far from clear (Wai, 2013). Regardless of whether there are sex differences in major psychological traits, men and women still opt for differential careers, even when there are similar levels of cognitive abilities.

## 9.2 Sex Differences in Science, Technology, Engineering, and Mathematics (STEM)

Why do women remain underrepresented in science, technology, engineering, and mathematics (STEM)? Are there sex differences in technical aptitudes that drive more males than females towards engaging in STEM careers? Is the attempt to strengthen young women's STEM-related abilities productive? Is the underrepresentation of women in STEM analogous to the underrepresentation of women in chess? These unknowns are noteworthy because they underlie a number of questions that have crucial implications within modern society, as reviewed in depth elsewhere (Halpern et al., 2007). A good deal of work addresses the sex gap in STEM fields, confronting at times radically opposing views. For example, it is disputed that men are more focused on objects, have a greater innate predisposition for mathematics and scientific fields, and are more variable in cognitive abilities in a review about sex differences in terms of intrinsic aptitude for mathematics and science (Spelke, 2005). This point of view aligns with the gender similarities hypothesis touched on earlier. In another view, there are consistent sex differences in mathematical reasoning ability that favour men. These differences increase at higher levels of ability, are stable over time, and are observed in several cultures (Benbow, 1988).

Further illustrating this ongoing controversy, two meta-analytic reviews into mathematical performance reach remarkably asymmetrical findings. The first meta-analysis addressed sex differences in mathematics achievement, with data from 242 empirical studies published between 1990 and 2007 (Lindberg et al., 2010). The main findings indicate robust evidence for sex similarities in mathematics performance, and that nationality, ethnicity, and age might potentially moderate the reported meta-analytic effect sizes. The second meta-analysis also addressed sex differences in mathematics and science achievement, with data from the United States' National Assessment of Educational Progress (NAEP) between 1990 and 2011 (Reilly, Neumann, & Andrews, 2015). In sharp contrast with the meta-analysis reporting sex similarities, this study reports small but stable mean sex differences in twelfth-grade mathematics and science, favouring males, within the study period. Moreover, these findings support the view that there is a large sex difference for high achievers at the upper tail of the skill distribution, with an overrepresentation of males by a 2:1 ratio in both mathematics and science. An additional extensive meta-analysis looked at sex differences in scholastic achievement, with 369 samples ranging from elementary to university educational levels and a wide time span from 1914 to 2011 (Voyer & Voyer, 2014). This study reports that females have an advantage in terms of the overall assortment of effect sizes, which is larger for language areas and smaller for mathematics areas. This advantage in the academic achievement of girls over boys is deemed to be highly stable considering the large time span of nearly a century. These latter findings and associated claims are somehow paradoxical, however, considering the underrepresentation of women in STEM fields and related professions.

Recent findings have contended that the sex gap in STEM is due to social inequalities (Else-Quest, Hyde, & Linn, 2010; Voyer & Voyer, 2014) or gender-science stereotypes (Nosek et al., 2009). On the other hand, other studies maintain that the sex gap may be caused by the complex interplay of multidimensional factors, while rejecting the influence of social inequalities and stereotype-related factors. From an evolutionary history approach, an earlier comprehensive explanatory model suggested that, because of sexual selection, men have a particularly developed neurocognitive system supportive of spatial cognitive ability. Such a system would explain sex differences in mathematical fields requiring skills apart from arithmetic such as geometry (Geary, 1996, 1999; Lubinski, 2010). In accordance with this model, sex differences in social styles and interests lead to greater male engagement in mathematics compared with women. From this same evolutionary point of view, it has been suggested that the sex differences in temperament and cognitive abilities mediated by sex hormones would induce sex differences in occupational behaviour (Browne, 2006).

Other comprehensive reports have shown that sex differences favouring men in the top-right tail distribution of mathematical ability have decreased substantially since 1981 (Wai et al., 2010). The most appropriate explanations for this trend should consider the simultaneous influences of multiple biological and socio-cultural factors rather than single-factor explanations. For example, stereotype effects and sex roles have been regarded as weak on the grounds of biological or biosocial models (Feingold, 1996). In particular, stereotype threat has been deemed unlikely to explain sex differences in STEM fields (Ceci, Williams, & Barnett, 2009). Only a rather weak effect is reported in a review of the available evidence in support of the stereotype threat, suggesting in fact that overemphasizing stereotype-related explanations might actually hamper effective interventions to close the gap (Stoet & Geary, 2012). In contrast to several claims arguing for discriminatory practices, the underrepresentation of women in science has been attributed to factors associated with family formation and childrearing, which affect women in all fields. In addition, career preferences and cognitive ability differences affect women particularly in the mathematics field (Ceci & Williams, 2011). Furthermore, a cross-national study with data from the Programme for International Student Assessment (PISA) demonstrates that national gender equality measures and sex differences in mathematics or overall achievement are rather unconnected (Stoet & Geary, 2015).

The wealth of findings from the Study of Mathematically Precocious Youth have been particularly informative in this regard. For example, it is somehow shocking that longitudinally stable sex differences emerge when comparing mathematically gifted men and women in dimensions of life other than abilities, such as occupational or life preferences, or opportunity (Benbow et al., 2000; Lubinski, Benbow, & Kell, 2014). Men commit to a greater extent than women do to developing impactful careers and invest more time in this facet of life. Women commit to a greater extent than men do to advancing family and community bonds. The SMPY findings also indicate that, among mathematically precocious people, more males than females persist in pursuing careers in STEM, whereas more females than males persist in pursuing careers in administration, law, medicine, and social sciences. These findings might be partly explained because women with high mathematical ability also have higher verbal ability than men, which might be additionally stimulating to seek careers other than in STEM fields (Wang, Eccles, & Kenny, 2013). The theory of work adjustment (TWA) is a useful general framework for understanding these patterns of sex differences in both educational and work scenarios (Lubinski & Benbow, 2006).

Apart from cognitive ability, another of the personal attributes that constitute the TWA conceptualization is the hexagonal interests model, incorporating Dale Prediger's two-dimensional structure, which comprises people–things and data–ideas axes (Holland, 1996; Prediger, 1982). In accordance with these latter two dimensions, sex differences in vocational and STEM interests emerge distinctly in a comprehensive meta-analytic review (Su, Rounds, & Armstrong, 2009). Across age and over time, men are more inclined to work with things, whereas women are more inclined to work with people. Moreover, the lower number of women in science and engineering fields compared with men is likely to follow from the predilection of women for careers focused on working with people. Is this predilection extensible to the chess domain?

## 9.3 Sex Differences in Participation Rates in Chess

There are fewer women in STEM fields and also in chess. There are, however, women at the grandmaster level who play very strong chess, just as men do. Nonetheless, when reviewing male to female ratios in earlier chapters (Table 5.1, Table 6.3, Appendix 3, Appendix 4), it becomes evident that women are markedly underrepresented in chess. Why is this so? Sex differences in cognitive abilities, personality, and interests have been extensively considered in terms of explaining the discrepancy regarding the amount of men and women involved in chess. For example, it has been argued that there is higher variability in the functional brain asymmetry during the menstrual cycle (Chabris & Hamilton, 1992). Physiological changes during the later phases of the menstrual cycle might contribute to decreasing right hemisphere activity, disturbing chunking abilities in highly demanding competitive chess tournaments, and hampering women's chess performance to a greater extent. At the physiological plane, and regarding the main male sex hormone, testosterone levels may also be helpful for chess performance. Greater concentrations of testosterone appear to buttress behaviours potentially suitable for winning chess games (Mazur et al., 1992). It has also been argued that sex differences in visuospatial abilities might account in part for the underrepresentation of women in chess (Frydman & Lynn, 1992). On the other hand, an extensive study into the personality of children ($n$ = 269) indicates that boys scoring lower in agreeableness are more prone to become involved in chess playing (Bilalić et al., 2007b). Agreeableness conveys behavioural traits implying trust, altruism, modesty, and tender-mindedness, with lower scores implying higher levels of aggressiveness, which might tend to bestow a certain advantage in a highly competitive domain such as chess. Because men tend to score higher in aggression-related psychological traits and manifest higher levels of aggression in real life, this would convey an additional advantage for males regarding chess performance (Anderson & Bushman, 2002; Hyde, 2005). Furthermore, sex stereotype mechanisms might be particularly influential for women in chess, as they appear to exert a meaningful impairment of chess performance for women (Maass, D'Ettole, & Cadinu, 2008; Rothgerber & Wolsiefer, 2014).

Because the Elo rating provides an objective measure of chess performance, several studies have addressed the different participation rates of men and women in chess by comparing the Elo ratings of men and women. One of the first studies in this line of research suggests a statistical function that is highly predictive of the observed differences (Charness & Gerchak, 1996). One of the main conclusions of this study is that participation rates should be taken into account when attempting to explain the differential chess performance of men and women. A subsequent study analysed a large sample of chess players from the United States Chess Federation ($n$ = 256,741) across thirteen years, from 1992 to 2004 (Chabris & Glickman, 2006). The main findings confirm higher mean Elo ratings for men than for women. Unlike the outcomes in several cognitive abilities, however, the Elo rating was more variable for females than for males. Besides, boys began to play competitive chess earlier than girls, particularly at higher performance levels. Thus, the main conclusion of this study revolves around the idea of a greater number of boys beginning to play chess at the lowest levels of skill. In addition, there are retrospective studies with Elo ratings data comparing several large samples of men and women chess players (Howard, 2005a, 2014a, 2014b). These studies report a robust and stable sex difference at the top ten and fifty of international players. The main argument of these studies highlights the fact that this disproportionate representation of male players at the top level of chess skill, together with the generalized male predominance in the chess domain, might be due to sex differences in innate abilities useful to developing chess skill.

Furthermore, the disparity in chess performance between men and women might depend on the differential participation rates of men compared with women, and with sex differences in the amount of chess practice, motivation, or interests (Bilalić & McLeod, 2006, 2007). A study with a large sample of German chess players ($n = 120{,}399$) compared the best 100 men with the best 100 women (Bilalić, Smallbone, et al., 2009). Sex differences in chess performance between these elite groups were rationalized as coming up because extreme values in a larger sample tend to be higher than extreme values in a smaller sample. When controlling for the remarkable disparity in the participation rates of men and women, no other factors – i.e., biological, cultural – are deemed to have been influencing the sex differences in performance. In contrast, a reanalysis with the same data from the German chess players with another statistical approach yields contradictory findings (Knapp, 2010). In the study by Bilalić and collaborators 96% of the observed differences between the best 100 men and the best 100 women are attributed to different participation rates. The study by Michael Knapp produces a sharply different figure, with 66% of these differences being attributed to different participation rates. These latter outcomes are argued to contravene the different participation rates hypotheses, as opposed to the contribution of biological or cultural factors to explain the sex gap in chess performance.

Later studies applying the statistical method suggested by Knapp also run counter to a strict differential participation rates hypothesis for explaining the sex gap in chess performance. For example, data from six chess tournaments in Spain included 511 males and 42 females, a male to female ratio of 12:1 (Blanch et al., 2015). This study evaluates whether participation rates or age and practice are able to account to a greater extent for the observed Elo rating differences between males and females. The main findings suggest that sex differences in the Elo ratings do not fully depend on the discrepancy in the participation rates across sex. Age and chess practice explain a substantial 69% of the variability in the Elo rating sex differences for females, while for males they explain only a scanty 8%. Another study examined a comprehensive sample of chess players (102,774 men and 9,585 women) from twenty-four countries in a cross-cultural comparative study (Blanch, 2016). This study applied the alternative method for looking at the problem of sex differences in participation rates in the chess domain (Knapp, 2010). There were sex differences in all the analysed countries, with meaningfully higher Elo ratings for men than for women in most countries. Moreover, the discrepancy in the participation rates of men and women were unable to fully explain the sex differences in chess performance.

Consider, for instance, the plots in Figure 9.1, which represent the ranks of the best 100 female players in Azerbaijan, Poland, the Czech Republic, and

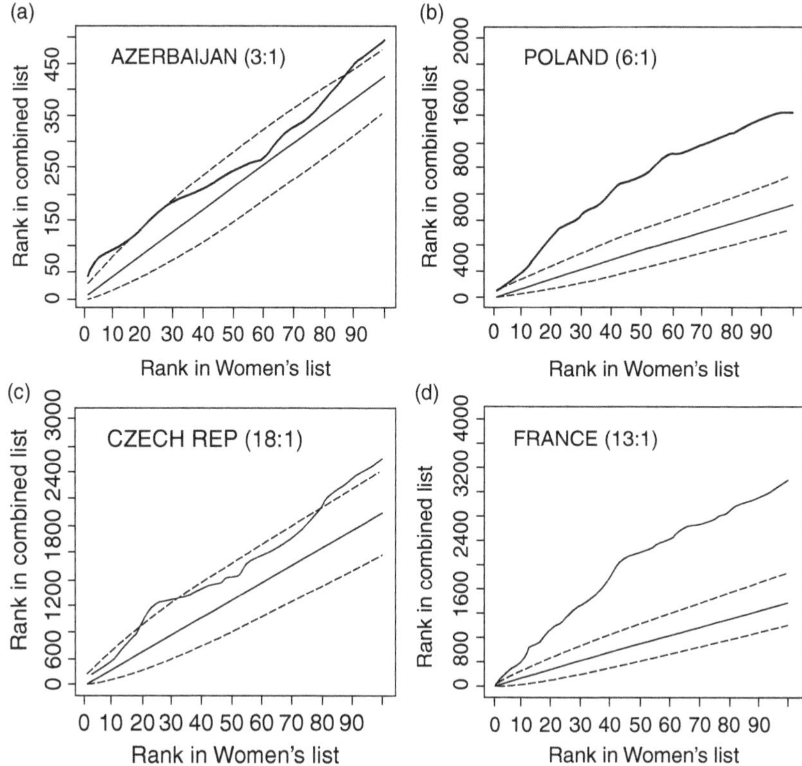


**Figure 9.1** Female observed ranks (black dots) compared with the expected rank (straight line), discontinuous lines representing the 0.05 and 99.95 quantiles; male to female ratios (M:F) are shown next to each country name

France. The X axis represents the rank of each player in the women's list, and the Y axis represents the rank in the combined male and female list. The observed ranks (represented by the black dots) falling inside the quantiles interval (dotted lines) support the view that sex differences in chess performance are due to the discrepancy in the participation rates, which is the case for Azerbaijan and the Czech Republic. In contrast, the observed ranks falling outside the quantiles interval indicate that sex differences in chess performance depend on factors other than the discrepancy in the participation rates, which is the case for Poland and France. The findings suggest that, regardless of the male to female ratio, sex differences in chess performance can be attributed either to the discrepancy in the participation rates of men and women (Azerbaijan and the Czech Republic) or to factors other than the discrepancy in the participation rates of men and women (Poland and

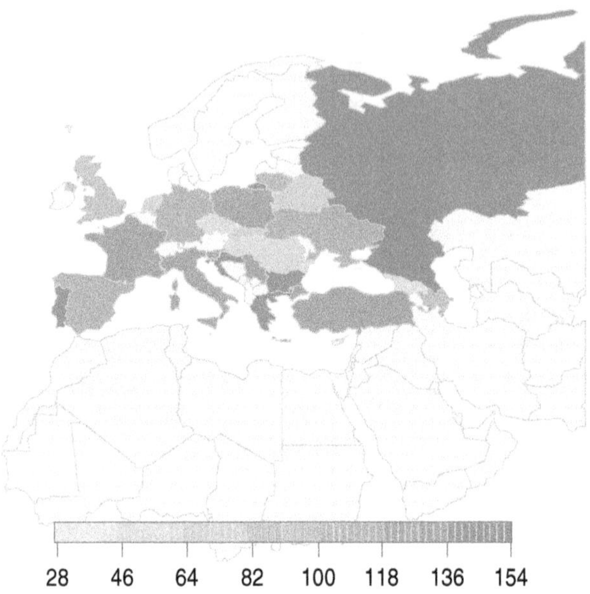


**Figure 9.2** Discrepancy in actual and estimated sex differences in Elo rating points for twenty-four Eurasian countries: Azerbaijan, Belarus, Bulgaria, Croatia, the Czech Republic, England, France, Georgia, Germany, Greece, Hungary, Italy, Lithuania, the Netherlands, Poland, Portugal, Romania, Russia, Serbia, Slovakia, Slovenia, Spain, Turkey, Ukraine

France). Hence, there was considerable variation across the twenty-four countries that did not systematically depend on the respective within-country male to female ratio. The map in Figure 9.2 shows the discrepancy in actual and estimated sex differences in chess performance for the twenty-four analysed countries. The discrepancies are higher for either larger countries or those closer to the Mediterranean Sea, whereas, in contrast, the discrepancies are lower for either smaller countries or those located in central Europe. What makes some countries have higher discrepancies than others is unknown and a worthwhile line of research.

The differences in the participation rates of men and women in chess are outstanding and somehow analogous to those found in STEM fields. Attributing sex differences to differential participation rates alone might be misleading, however. Once you have controlled for the participation rates of men and women, several factors might account for the observed differences in performance. Moreover, there is a crucial question that should be asked at this point: are there meaningful differences in chess playing between men and women?

## 9.4 Sex Differences in Chess Playing

Several factors encompassing individual differences are influential for chess performance, such as brain functioning, cognitive abilities, and dispositional traits such as personality, motivation, and emotional regulation. On the other hand, there are noteworthy gender differences in various of these factors. Therefore, the question may be raised as to whether men and women harness different chess-playing strategies. The issue has stimulated a number of interesting studies in this field.

Sex differences in strategic behaviour were compared in a retrospective study by analysing a large number of chess games (1.4 million) and chess players ($n = 15{,}000$), and by devising a classification of chess openings into two main styles: aggressive and solid (Gerdes & Gränsmark, 2010). Aggressive openings would imply higher risk and more attacking options, whereas solid openings would imply lower risk and fewer attacking options. Both kinds of opening styles reflect well the general combinative and positional chess-playing schemes suggested by Nikolai Krogius (Krogius, 1976), and the four main styles indicated by Fernando Peralta and Alejo de Dovitiis (Peralta & de Dovitiis, 2007) – i.e., ambitious and risky with regard to aggressive openings, and conformist and robust with regard to solid openings. In the view of the authors of the study it is further expected that the player's temperament might modulate the choice of specific playing styles – i.e., calm and peaceful individuals practising more often a solid opening style. Moreover, with individual differences in personality between men and women, one would expect differences between men and women when making use of either aggressive or solid opening strategies. These are, in fact, the kinds of sex differences reported here. Women endorsed more solid openings than men did. In addition, an unexpected and stunning finding is that both men and women endorsed aggressive openings and avoided solid openings when playing against a female opponent even when playing stronger opponents in terms of higher Elo ratings. This was the main choice even at the expense of reducing the probability of winning.

Further analyses of sex differences in chess playing look at impatient behaviour by considering the duration of games (Gränsmark, 2012). This study examined a large amount of games (1.5 million) with more than 30,000 chess players from 140 countries. Men leaned towards playing shorter games than women did. Men performed worse in shorter games than women did, but performed better in longer games compared with women. Women performed worse when exposed to time pressures, however, corroborating the higher impatience of men compared with women. In contrast, women appeared to invest a longer time in the opening stages of the game, eventually leading to time pressures later. A third study, from the same group of Swedish researchers, addresses the association of physical attractiveness and risk taking in chess, extending part of the previous findings with the same dataset (Dreber,

Gerdes, & Gränsmark, 2013). The key finding from this study is that male chess players selected higher-risk playing strategies when playing against a physically attractive female player, even though this course of action did not improve their chess performance. In contrast, women's chess-playing strategies appeared to be unaffected by the physical attractiveness of male opponents.

Sex differences in chess have also been looked into by studying chess-related stereotypes. A stereotype is a generalization of a belief for all the individuals in a particular category of people. When the question 'Who is better at playing chess: men or women?', plus three additional related questions, were put to a group of twelve male chess players and ten female chess players, it was found that both men and women subscribed to the view that males are better and more gifted at chess than females (Maass et al., 2008). In this study, an additional experimental set-up matched forty-two male chess players with forty-two female chess players of an equivalent Elo rating skill. Each dyad played two online rapid games with fifteen minutes per player within either control or experimental conditions. In the control condition, both players were aware that they were playing against an opponent with a similar chess strength. In the experimental condition, participants were told about the better outcomes of men over women in chess. In addition, participants were told that they were playing against a same-sex opponent in one of the games, and that they were playing against an opposite-sex opponent in the remaining game. The findings were staggering. In the control condition, without information about the sex of the opponent, women won nearly half the games. In the experimental condition, with information about the same-sex opponent, women's performance was analogous to the control condition. The performance of women with information about the opposite-sex opponent dropped by 50%, however. It is suggested that these findings support the primary role of gender stereotypes in chess, with female players being more intimidated when they thought they confronted a male player, and adopting a more defensive strategy. Moreover, women played as well as their male opponents when they thought they were confronting a female player.

Chess-related stereotypes appear to emerge also in childhood. Another cross-sectional study with a more extensive sample compared a group of girls ($n = 219$) and boys ($n = 195$), from kindergarten to ninth grade (approximately from five to fifteen years old), regularly involved in competitive chess-playing tournaments (Rothgerber & Wolsiefer, 2014). The findings show worse female performance when playing male opponents, with even stronger effects at younger ages. Moreover, the authors argue that the stereotype threat for female players was particularly salient in more stringent circumstances when playing either older or moderate to strong opponents, whereas these findings were not replicated with the male players. In stark contrast with these latter findings, however, a more recent study about the gender stereotype threat in chess reaches radically distinct conclusions (Stafford, 2018). The analysis in this study encompassed over 9 million chess games, corresponding to a very large sample of chess players ($n = 461{,}637$, with 56,474 females). The findings certainly suggest that the chess performance of female players is largely unaffected by gender-stereotyped constraints when playing against male opponents. This was also the case even when male opponents were much stronger in accordance with their respective Elo ratings.

A certain level of aggressiveness may be useful in a warfare-like game such as chess. The aggressive component has been observed with children recording lower scores in the agreeableness personality factor. Those children more keen to become involved in chess reported lower scores in the agreeableness element of the big five broad personality factors, which may be an additional influential variable for entering and persisting in the chess domain (Bilalić et al., 2007b). Thus, it can be conjectured that the higher aggressiveness levels of men may be stimulative of their higher involvement and performance when compared with women. Furthermore, the study with elite German chess players that compared the personality of male and female players with the normal population did indeed yield meaningful sex differences that might support this standpoint about aggressiveness (Vollstädt-Klein et al., 2010). Although the male personality scores were analogous to those from the general population, female chess players scored higher in life satisfaction and achievement orientation, and lower in somatic complaints. The authors suggest that these findings support the view that high achievement motivation, toughness, and stability are requirements for succeeding in elite chess, which might be particularly influential for women when contending in such a male-dominated activity. From this point of view, women with a greater predisposition to personality traits more attuned to such a competitive field would be more likely to succeed within the domain.

## 9.5 Sex Differences in Chess Performance at Different Levels of Practice

The chess domain is particularly compelling for evaluating whether individual differences in performance are determined by practice or by factors other than practice. Men outperform women in chess, a domain requiring remarkable cognitive effort and high investment in practice activities to attain an expert level. This section examines sex differences in chess performance at different levels of practice with an extensive sample of male and female players (4,405 men and 600 women). The main aim of this analysis is to assess whether sex differences in chess performance change at different levels of practice.

Men achieve higher chess performance than women do at the expert level. Because women are underrepresented in chess, these sex differences in chess performance have been categorized as a statistical artefact associated with the different participation rates of men and women in the domain (Bilalić & McLeod, 2006; Bilalić, Smallbone, et al., 2009; Chabris & Glickman, 2006; Charness & Gerchak, 1996). There could be other influential factors underlying these differences, however. For instance, starting age, practice, cognitive abilities, and stereotype threat have all been repeatedly pointed to as potential predictors of the observed sex differences in chess performance (Blanch, 2016; Blanch et al., 2015; Chabris & Glickman, 2006; de Bruin et al., 2008; Howard, 2005a, 2014a, 2014b; Knapp, 2010; Maass et al., 2008). Moreover, the interrelationships between deliberate-practice-related factors might further enhance the impact on chess performance, such as starting serious play at early ages or birth season, group practice, specialized coaching, book reading, and persistence within the domain. Empirical evidence suggests, for instance, that practice and chess-training activities, including tournament competition, might be more profitable at earlier than at older ages (Campitelli & Gobet, 2008, 2011; Gobet & Campitelli, 2007; Roring & Charness, 2007).

In a domain such as chess, high degrees of study, effort, and practice are undoubtedly needed to reach an expert level. Estimations for reaching expert chess performance range from a minimum of 3,000 hours (Campitelli & Gobet, 2011) to an average of 10,000 hours of total practice (Simon & Chase, 1973). An in-depth account of practice within expertise research is provided by the deliberate practice approach, which considers a conglomerate of intentional and methodical undertakings thought to improve skills and accomplishments in a given domain. In accordance with this framework, people achieve expert performance only after a period of intense deliberate practice, extending from their introduction to the domain to either the attainment of professional level or dropping out (Ericsson, 2006; Ericsson et al., 1993). Nevertheless, this approximation has been questioned on several grounds, because it explicitly excludes alternative factors for expert performance (Ackerman, 2014; Ericsson et al., 1993; Hambrick et al., 2014; Howard, 2011a). Deliberate practice has been distinguished from actual practice. In the chess domain, coaching, reading books, studying chess openings or endgames, game analyses, and playing with a computer have been deemed to be deliberate practice activities. In contrast, actual practice comprises the competitive facet of practice within sports, which in chess implies playing chess games versus other human chess players (Howard, 2009). The amount of chess games appears to be a feasible and reliable indicator of actual chess practice – i.e., tournament activity (Blanch et al., 2015; Gobet & Campitelli, 2007; Howard, 2012a; Vaci et al., 2015).

A considerable degree of practice is a necessary but not a sufficient condition to achieve an expert chess level (Campitelli & Gobet, 2008, 2011). Because a considerable amount of practice is required to achieve chess expertise, practice might therefore contribute to a great extent to explaining sex differences in chess performance. Hence, it is pertinent to assess whether the overall pattern of sex differences in chess performance varies at diverging levels of chess practice. Earlier works about the effect of practice in performance suggested that not everybody benefits equally from comparable amounts and/or quality of practice activities. The inter-individual variability in performance increases with cumulative practice when factors other than practice are important, whereas the inter-individual variability in performance decreases with cumulative practice when practice in itself is important (Anastasi, 1937; Thorndike, 1908). This may depend as well on the kind of task (see Table 8.2), however. Individual differences after the impact of practice are likely to lessen in tasks involving speed and accuracy, whereas, in contrast, individual differences after the impact of practice are likely to grow in tasks involving domain knowledge (Ackerman, 2007).

Because chess involves a considerable knowledge base, decrements in individual differences in chess performance with cumulative practice could be attributable to practice, whereas increments in individual differences in chess performance with cumulative practice could be attributable to factors other than practice. Therefore, lower sex differences in chess performance at increasing levels of practice should support the importance of chess practice for explaining sex differences in chess performance. On the other hand, higher sex differences in chess performance at increasing levels of practice should support the importance of factors other than practice for explaining sex differences in chess performance. These hypotheses were addressed by examining an extensive group of men and women chess players. Information about the amount of games as the practice indicator and the Elo rating as the chess performance indicator (Elo, 1978) was collected from active chess players. Sex differences in chess performance according to the Elo rating were compared between the men and women players by analysing the pattern of change in sex differences at varying levels of practice.

Data from the Elo rating list published by the World Chess Federation in March 2014 was the input for the current study (www.fide.com): the Elo ratings, the amount of games, and the age of over 173,000 men and women players from 170 countries. Only active players who had age data and had played at least one game within the period were included in the present analyses, which left 4,405 men and 600 women, with a men to women ratio of 7:1. The mean number of chess games of men and women were significantly different ($t = 2.30$, $p < 0.05$), with a mean of 7.65 games for men ($Sd = 4.96$) and 7.25 games for women ($Sd = 3.97$). Moreover, the mean age for men and women was 33.92 ($Sd = 17.64$) and 22.32 ($Sd = 11.19$), respectively, with a statistically significant difference ($t = 21.95$, $p < 0.001$). Mean Elo ratings for men and women were also significantly different ($t = 14.91$, $p < 0.001$), with a mean Elo rating for men and women of 1987 ($Sd = 314$), and 1788 ($Sd = 305$), respectively. For men,

Pearson's correlations between these variables were –0.09 for games with age, 0.18 for games with the Elo rating ($p < 0.001$), and 0.15 for the Elo rating with age ($p < 0.001$). For women, Pearson's correlations were –0.09 for games with age ($p < 0.05$), .23 for games with the Elo rating ($p < 0.001$), and 0.37 for the Elo rating with age ($p < 0.001$).

Sex differences in chess performance were examined by selecting men and women with an equal level of practice in accordance with four practice layers. In each layer there were twelve, six, three, and two levels, indicating a growing amount of games played in the period. There were twelve different levels with fifty men and fifty women in the first layer, six levels of 100 men and 100 women in the second layer, three levels of 200 men and 200 women in the third layer, and two levels of 300 men and 300 women in the fourth layer. While women remained the same across the different game levels ($n = 600$), an equivalent amount of men were selected at each level matching the number of games for women.

Table 9.2 shows an overview of the mean games in each level within the four layers. The next column shows the mean Elo ratings for men and women at each level. This allows us to examine sex differences in the Elo rating at a varied spectrum of practice levels. For example, and for level number 2 within layer 1, the fifty men and fifty women players all had the same mean number of games, 3.02 games, with a standard deviation for men and women of 0.14 and 0.55, respectively. It can be seen that, for this particular instance, however, the mean Elo rating for men was 2258 (227), whereas the mean Elo rating for women was 1594 (397), a difference that was statistically significant ($t = 10.28$, $p < 0.001$).

The mean Elo rating differences between men and women were examined with an analysis of variance (ANOVA) with the Elo rating as the dependent variable, and sex, age, and games as the independent variables, together with the interaction terms 'Sex × games' and 'Sex × games × age'. Tukey's honest significant difference post-hoc analyses were also performed, to evaluate whether mean Elo ratings for men and women were significantly different at each level within each practice layer (Tukey, 1949). This approach is appropriate for a balanced design with the same observations at each level of practice. All data analyses were performed with the R software (R Development Core Team, 2015).

The analyses of variance with the Elo rating as the dependent variable yielded highly significant $F$ statistic values for the three independent variables – sex, practice, and age – within each of the four layers ($p < 0.001$). The last two columns in Table 9.2 show the Tukey post-hoc analyses of variance outcomes for the interaction terms of 'Sex × games' and 'Sex × games × age'. The entries at each level indicate the differences between men and women in mean Elo ratings. For both interaction terms, the $F$ statistic was significant at the four layers. Across the four analysed layers, the mean differences in the Elo ratings were particularly significant at the extreme levels. Consider, for instance, layer 1, with

Table 9.2 *Means, standard deviations (Sd), and* t-*tests in number of games, and Elo ratings for 600 men and women chess players. Layer 1 has twelve levels with fifty men and fifty women; layer 2 has six levels with 100 men and 100 women; layer 3 has three levels of 200 men and 200 women; layer 4 has two levels of 300 men and 300 women*

| | Games | Elo rating | | | | Tukey post-hoc analyses | |
|---|---|---|---|---|---|---|---|
| Layer | Men and women M (Sd ♂, Sd ♀) | Men M (Sd) | Women M (Sd) | $t$ | | Sex × games | Sex × games × age |
| **Layer 1** | | | | | | | |
| 1 | 1.44 (0.50, 0.50) | 2006 (594) | 1767 (293) | 2.55* | | 239* | 280*** |
| 2 | 3.02 (0.14, 0.55) | 2258 (227) | 1594 (397) | 10.28*** | | 664*** | 645*** |
| 3 | 4.06 (0.24, 0.24) | 2115 (271) | 1656 (253) | 8.76*** | | 459*** | 392*** |
| 4 | 5 (0, 0) | 2273 (142) | 1618 (176) | 20.49*** | | 655*** | 585*** |
| 5 | 5.82 (0.39, 0.39) | 1620 (433) | 1697 (207) | −1.14 | | −78 | −4 |
| 6 | 6.66 (0.48, 0.48) | 1782 (502) | 1725 (248) | 0.71 | | 56 | 102 |
| 7 | 7.36 (0.48, 0.48) | 2121 (561) | 1808 (286) | 3.51*** | | 313*** | 309*** |
| 8 | 8.06 (0.24, 0.24) | 2335 (309) | 1880 (218) | 8.53*** | | 456*** | 446*** |
| 9 | 9 (0, 0) | 2482 (25) | 1780 (171) | 28.79*** | | 703*** | 652*** |
| 10 | 9 (0, 0) | 2594 (38) | 2166 (131) | 22.19*** | | 428*** | 462*** |
| 11 | 10.84 (0.37, 1.17) | 1840 (378) | 1839 (371) | 0.01 | | 0.98 | 55 |
| 12 | 16.68 (0.47, 2.9) | 2125 (267) | 1927 (329) | 3.31*** | | 198 | 169 |
| | | | | | ***F*** | **17.07***** | **9.69***** |

Table 9.2 *Cont.*

| | Games | Elo rating | | | | Tukey post-hoc analyses | |
|---|---|---|---|---|---|---|---|
| Layer | Men and women M (Sd ♂, Sd ♀) | Men M (Sd) | Women M (Sd) | *t* | | Sex × games | Sex × games × age |
| **Layer 2** | | | | | | | |
| 1 | 2.22 (0.42, 0.95) | 2131 (479) | 1680 (358) | 7.52*** | | 450*** | 459*** |
| 2 | 4.53 (0.50, 0.50) | 1754 (415) | 1637 (217) | 2.50* | | 117 | 105 |
| 3 | 6.24 (0.43, 0.61) | 2052 (388) | 1711 (227) | 7.57*** | | 341*** | 333*** |
| 4 | 7.71 (0.46, 0.52) | 1862 (453) | 1844 (256) | 0.35 | | 18 | 43 |
| 5 | 9 (0, 0) | 2491 (55) | 1973 (246) | 20.54*** | | 518*** | 513*** |
| 6 | 13.76 (0.53, 3.67) | 2087 (386) | 1883 (351) | 3.90*** | | 203** | 195*** |
| | | | | | ***F*** | **18.82*** ** | **10.75*** ** |
| **Layer 3** | | | | | | | |
| 1 | 3.38 (0.49, 1.38) | 1784 (357) | 1659 (296) | 3.83*** | | 126*** | 122*** |
| 2 | 6.97 (0.16, .93) | 1706 (202) | 1778 (250) | −3.14** | | −71 | −71 |
| 3 | 11.38 (0.62, 3.52) | 2100 (333) | 1928 (306) | 5.38*** | | 172*** | 175*** |
| | | | | | ***F*** | **22.03*** ** | **9.09*** ** |
| **Layer 4** | | | | | | | |
| 1 | 4.33 (0.47, 1.79) | 1757 (273) | 1676 (276) | 3.59*** | | 80** | 74** |
| 2 | 10.16 (0.67, 3.37) | 2088 (350) | 1900 (293) | 7.15*** | | 188*** | 195*** |
| | | | | | ***F*** | **13.43*** ** | **4.73* ** |

*Notes:* * $p < 0.05$; ** $p < 0.01$; *** $p < 0.001$.

twelve levels, where the differences in Elo ratings were irrelevant at the intermediate fifth and sixth levels, and at the highest eleventh and twelfth practice levels. In both interactions ('Sex × games', 'Sex × games × age') the highest mean difference in the Elo ratings for men and women were generally found at high practice levels. For instance, there were 703 and 652 Elo points at the ninth level in layer 1, 518 and 513 Elo points at the fifth level in layer 2, 172 and 175 Elo points at the third level in layer 3, and 188 and 195 Elo points at the second level in layer 4.

Figure 9.3 describes the Elo ratings for the 600 men and women players at the four practice layers in twelve levels (first plot), six levels (second plot), three levels (third plot), and two levels (last plot). For the first layer, there is a marked gap between men and women in levels 1, 2, 3, and 4, dropping at levels 5 and 6, and increasing progressively from levels 7 to 10. For layers 2 and 3 there is also a similar trend. There are higher differences in the Elo ratings at the extreme groups that equalize at the middle groups. The findings show that sex differences in chess performance increase as the level of practice increases, suggesting that factors other than practice could account for the observed differences.

The aim of the current analysis was to assess whether sex differences in chess performance change at different levels of practice. The number of games was strongly correlated with chess performance for men and women players, corroborating the importance of actual practice in terms of tournament activity – an essential component for reaching expert performance in the domain. Sex differences in chess performance, as indicated by the Elo rating, increase progressively at growing levels of tournament activity. Taken together, these outcomes suggest that, despite the fact that practice is a determining factor for achieving expert performance in chess, factors other than practice are also likely to influence the observed disparity in chess performance between men and women.

Past research has highlighted the importance of practice in accounting for mean national Elo ratings (Campitelli & Gobet, 2008). In the current analysis, there were significant differences for the whole sample, even though the amount of games was very similar on average for both men (7.65) and women (7.25). By defining different levels of equivalent practice for men and women, the current assessment reveals remarkable intergroup differences between men and women in chess performance, which increase significantly at growing levels of practice. These findings characterize higher sex differences with cumulative practice, and suggest that factors other than practice are important in determining sex differences in chess performance (Anastasi, 1937; Thorndike, 1908). Moreover, the findings agree with what should be expected regarding skilled performance for a task such as chess, for which a large knowledge base is required (Ackerman, 2007). From this point of view, potential determinants of sex differences in chess performance could be narrower traits embedded within crystallized abilities, interests, personality,

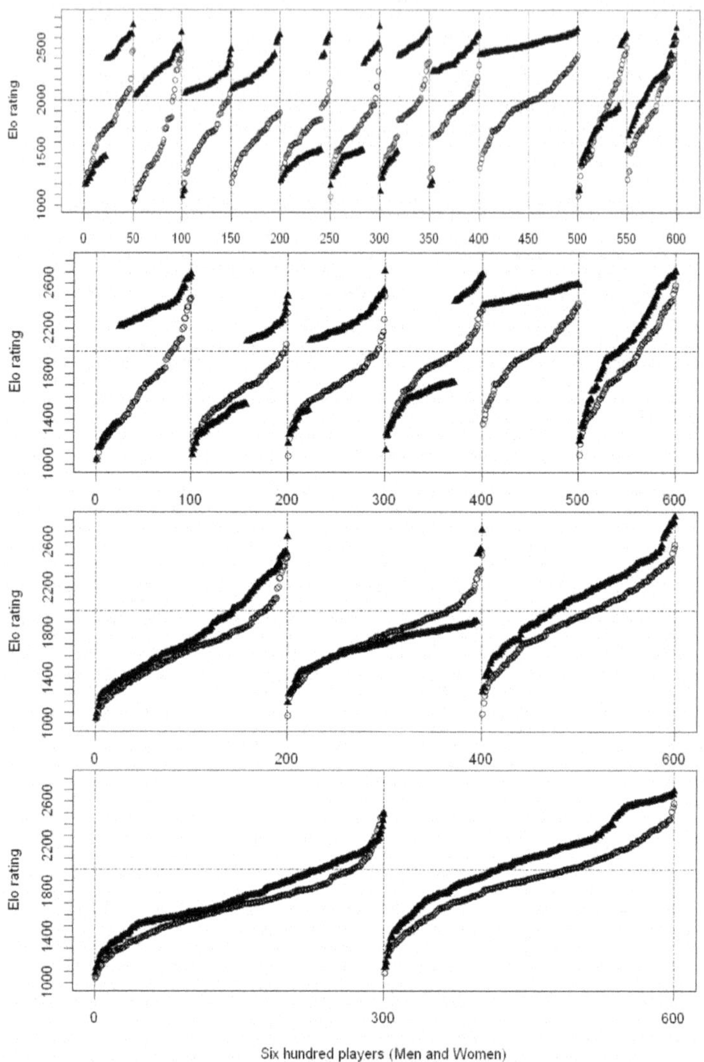


**Figure 9.3** Elo ratings for men (black triangles) and women (white circles) at twelve, six, three, and two levels of practice; the horizontal dotted line represents 2000 Elo points

and motivation (see Table 8.2). Indeed, the findings suggest that practice is not the key determining explanatory factor of sex differences in chess performance.

A direct apparent implication of this outcome is that sex differences in chess performance are independent of the fact that, on average, men tend to play more games than women do (Blanch et al., 2015; Campitelli & Gobet, 2008, 2011). For instance, women might play a lower amount of games than men because they become uninterested and inactive in the domain sooner than men (Howard, 2005a, 2014a, 2014b). Moreover, with fewer women enrolling at the lowest rating levels, their lower levels of practice, and higher dropout rates, those women persisting in the domain may be particularly motivated to play and do well in chess tournaments (Chabris & Glickman, 2006; de Bruin et al., 2008; Howard, 2005a, 2009; Roring & Charness, 2007). A variety of factors proposed in past research may contribute therefore to elicit sex differences in chess performance, such as starting age, other forms of practice aside from the number of games, handedness, cognitive abilities, and social factors (Campitelli & Gobet, 2011; Gobet & Campitelli, 2007). At a broader level of analysis, it is likely that the combination of cognitive abilities and social variables, distally determined by the interaction of genetic and environmental factors, could also account for the great disparity in chess performance of men and women (Blanch, 2016; Chabris & Glickman, 2006; de Bruin et al., 2008; Howard, 2005a; Maass et al., 2008).

The Elo rating correlates with constructs such as cognitive abilities and personality, which convey noteworthy sex differences in applied fields (Blanch & Aluja, 2013; Grabner, 2014a; Vollstädt-Klein et al., 2010). Even though spatial abilities have been considered negligible for the progression of chess skill in adults (Waters et al., 2002), it has also been argued that the higher performance of men in this set of abilities within the chess domain could influence sex differences in chess performance (Frydman & Lynn, 1992; Geary, 1995a; Horgan & Morgan, 1990). Mnemonic abilities are also considered important for chess performance, as conceptualized by the template theory (Gobet, 1998). Future research works addressing the association of sex differences in chess performance with spatial, and particularly memory-related abilities, might be worthwhile because women tend to obtaining higher average scores than men in tasks requiring rapid access and retrieval of information stored in long-term memory (Halpern, 1997).

Furthermore, some review studies have argued for the key influence of social pressures to explain the preponderance and better performance of men over women in quantitative and scientific areas (Halpern, 1997; Steele, 1997), while arguing for equivalent innate aptitudes of males and females (Halpern & Wright, 1996; Spelke, 2005). With a similar biological endowment for males and females concerning cognitive abilities, the marked sex differences in chess performance might therefore be attributable to environmental influences spreading towards a deeper and more intense involvement of men than women in the domain. The standpoint considering both biological and environmental influences in the inception and development of differential sex cognitive abilities has gained a favourable consensus within the field (Halpern, 1996; Lubinski, Benbow, et al., 2001; Maass et al., 2008; Steele, 1997). For instance, the significance of hormonal variations has been highlighted as a proximal biological cause in shaping sex cognitive configurations, even though their breadth and expression are notably modulated by contextual and cultural factors (Benbow, 1988; Geary, 1995a, 1995b).

These outcomes parallel those concerning the considerable body of research into sex differences in several psychological traits. Factors other than practice influence the extreme differences in chess performance of men and women. Studying sex differences in chess performance at varying levels of practice is important because practice is a key central factor for chess expertise development. Nevertheless, these outcomes indicate that actual chess practice, as calibrated by the amount of games played, is unable to fully explain the observed sex differences in chess performance. The differences in chess performance between men and women could be explained by the variety of biosocial factors described so far.

# 10

# Applications

The game of chess is universal. It has only a few rules, and they can be learned effortlessly at diverse age and educational levels. Besides, it is relatively easy and cheap to obtain a chess set, and the available chess information and applications for playing chess online through information technology have grown exponentially. These factors have probably contributed to the use of chess on applied fields with different practical purposes. Chess has been applied in three main broad applied domains: business, health, and education. The studies that have been conducted in each of these three domains by using chess in some way or other are described in this chapter.

## 10.1 Business

Chess is regarded as providing several educational benefits in business academic studies. For example, it has been argued that stimulating chess playing among business students may help illuminate several abstract concepts during their formal education, and later on for practical purposes in their professional careers (Graber, 2009). There are several useful parallels and analogies between chess and business, such as the plus-sum game nature, ethical issues, long-term thinking, investment strategies (i.e., chess gambits), flexibility in designing contingency plans, patience and perseverance, and teamwork. One specific exercise in this field implemented a simulation intervention with 113 undergraduate students and seventeen graduate students in business management to introduce or illustrate key management concepts (Cannice, 2013). This intervention was considered useful and rewarding for students as an experiential tool for consolidating management concepts learned in regular lecturing and case discussions. Moreover, the self-explanation effect with chess material was studied in an experimental study with forty-five undergraduate psychology students (de Bruin, Rikers, & Schmidt, 2007a). The self-explanation effect claims that individual differences in the ability to self-explain problem-solving strategies are beneficial for learning. An interesting implication for education and instruction derived from this study concerns the verbalization and explanation of reasons that are eventually conducted for finding solution steps.

Chess data have also been used in the more exhaustive context of game theory. For example, the centipede game has been undertaken in two studies with considerable sample sizes of chess players, albeit with contradictory findings. The centipede game is a finite-move game involving two players. The game consists of choosing alternately, between the two players, whether to finish or to pass to the other player. There is a payoff received by the player who ends the game, which is greater than the payoff if the other player ends the game, but which is lower when the other player continues with the game. The best strategy in this game is to stop at the first node (i.e., equilibrium), an outcome rarely found in empirical studies.

A study with 422 Spanish chess players argues in favour of the usefulness of applying this game in an experimental set-up with chess players, because of their high degree of rationality and familiarity with backward induction (Palacios-Huerta & Volij, 2009). Players who were more skilled stopped more frequently at the first node of the game. The implied stop probability at the two first nodes decreased progressively with chess skill, from the grandmaster level to average chess players. An additional experiment, within the same study, paired undergraduate students with no idea about chess with chess players. When students played against each other, very few students stopped at the initial node (3%), while a considerable amount stopped at nodes 3 or 4 (60%). This distribution changed dramatically, however, when students were informed (player 1) that they were playing against chess players, with an increase in the observations stopping in node 1 by a factor of ten (30%). This finding suggests that students modified their behaviour depending on their beliefs about the rationality of their opponents. There are two main conclusions. First, chess players behaved very differently from students in the centipede game. Second, the majority of chess players chose the only action consistent with equilibrium. A subsequent similar study with the centipede game and chess players ($n = 206$) from the United States challenged somehow the previous findings with Spanish chess players, however (Levitt, List, & Sadoff, 2011). The centipede findings resembled more closely the empirical findings from the general population. For example, in the pairings (sixteen) with a chess grandmaster as player 1, none of them stopped at the first node. Moreover, chess skill was unrelated to stopping at earlier nodes, even though grandmasters, international masters, and other masters stopped more frequently at nodes 3, 4, and 5. These findings were attributed in part to the cooperative behaviour that occasionally arises in chess, such as when both players agree to a draw in the course of a game.

Another study looking at such cooperative behaviour in chess tested the null hypothesis that chess players strive to win a given chess game or tournament. Here, the alternative hypothesis was the potential collusion of Soviet players aimed at drawing chess games when playing other Soviet players, in order to maximize the likelihood of a Soviet winning the tournament (Moul & Nye,

2009). Over 7,000 chess games played from 1940 to 1978 were examined to analyse the draw chess outcome and draw-like scenarios in real life. The main finding in this study suggests that the Soviets were indeed more likely to draw chess games when playing among themselves within the observed period. This behaviour implied a substantial modification of chess playing oriented towards balanced, conservative, and low-risk strategies. The authors wonder whether similar behaviour encouraging 'draws' could be extrapolated to business applications, such as the structuring of incentives in strategic competitions, or the designing of pay packages in managerial hierarchies.

The three aforementioned studies from Sweden about sex differences in chess also adopted econometric models by scrutinizing a large assortment of chess games with sizeable samples of chess players. Apart from providing several compelling views about sex differences in chess, the findings from these studies have also been interpreted in the light of strategic behaviour in negotiation and economic events. For example, the findings contrasting aggressive versus solid openings for men and women corroborate the higher aversion of women compared to men when making risky decisions (Gerdes & Gränsmark, 2010). In addition, these findings suggest that, in negotiating situations, women might tend to avoid competition, and fail when forced to do so. This behaviour is clearly distinguished from the typical behaviour of men, who might be more overconfident, particularly in male-dominated realms such as trading and business. Furthermore, the economic interpretation of the findings from the study about impatience and sex differences emphasizes the greater impatience of men compared to women (Gränsmark, 2012). Men were more inconsistent because of being more impulsive, and women were more inconsistent because of reflecting too long. Moreover, the study relating physical attractiveness and risk taking reports that men adopted a risky playing style when playing against an attractive female opponent (Dreber et al., 2013). Here it is argued that these effects could be extrapolated to a variety of situations in the labour market. For example, men might take on higher risks when confronted with attractive women when playing in a casino, giving a loan, or choosing financial products.

Another extensive study with a large sample of over 40,000 chess players considered all the FIDE tournaments played in the world between 2008 and 2011 (Bertoni et al., 2015). Each tournament reported the results of all the games played by the participants in terms of wins, losses, or draws. The aim of the study was to examine the association of age with mental productivity. Individual differences in age-related declines in productivity might be strongly linked with overall productivity and bear fundamental macroeconomic implications. The main findings indicate that less talented players are more likely to drop out with ageing, that the effect of selection is more pronounced at earlier ages, and that chess experts who were forty years old were around 10% less productive than younger players who were fifteen years old. This econometric study suggests in addition that maturation in crystallized intelligence is unable to compensate for the early decline in fluid intelligence in professional chess. Overall, this study highlights the robust interrelationship between age, ability, and productivity in the light of an econometric approach. Further econometric work has also addressed the role of cognitive abilities. A study with over 6,000 chess players who underwent a game-theoretic experiment, named as a beauty contest, concludes that higher chess-playing skill is not associated with higher levels of rational choices (Bühren & Frank, 2012). In addition, the study underlines remarkable limitations concerning the transfer of cognitive abilities across different domains, and notable individual differences among the biographies of top chess players concerning their complementary talents other than chess playing (see Table 6.3).

## 10.2 Health

Chess has also been used to look at health problems, particularly in relation to mental and behavioural disorders. For instance, learned helplessness theory describes a set of impairing behavioural events that arise when the outcomes of a situation are perceived as uncontrollable. These events are associated with motivational, cognitive, and emotional deficits, eventually leading to clinical depression or severe mental disturbances (Abramson, Seligman, & Teasdale, 1978). An experimental study addressed learned helplessness with forty-eight Swiss chess players, by manipulating the solution in a set of fourteen chess problems and comparing two different tasks (Gobet, 1992). In a normal feedback group the chess problems encompassed an objective solution. In a learned helplessness experimental condition, however, the chess problems lacked an objective solution. The outcomes were compared across four categories of players in accordance with their Elo ratings: from 1600 to 1800 points in category IV, and from 2200 to 2450 points in category I. Subjects in the learned helplessness condition performed worse than their colleagues. Moreover, the players in categories IV and I were less sensitive to the learned helplessness condition than those in categories III and II were, albeit these differences were statistically meaningless. Another study used psychometric measures of helplessness/hopelessness from nine male active players above 2300 Elo rating points, which were compared with a cardiac response during championship games (Schwarz et al., 2003). Increased hopelessness associated with reduced high-frequency heart rate variability (HF-HRV), indicating that negative mood states associate with observable cardiac events. This is a beguiling finding, because the study situation – real-life chess games in a championship – was much more natural than inducing the affect situation in a laboratory setting. The main outcomes of the study, however, were that increased helplessness/hopelessness elicited concurrent increases in sympathetic tone and reflected a fight–flight condition, which might lead eventually to undesirable cardiac events such as ventricular fibrillation, bradycardia, or asystole.

Chess playing has been used to improve executive functions in patients with schizophrenia (Demily et al., 2009). This study selected only patients with a normal intellectual level, the absence of resistance to treatment or traumatic brain injury, and the presence of other neurologic or mental illness. The intervention lasted five weeks, with two chess teaching sessions per week, the chess training totalling ten hours. After the chess-training intervention, there were meaningful improvements in the Stroop and Tower of London tests, supporting the amelioration of voluntary processing, distraction inhibitory capacity, and planning abilities. It is furthermore remarkable that this study reports that the patients continued to play chess on their own after the completion of the study, highlighting the appeal of the game for people with a very severe psychiatric disorder such as schizophrenia. On the other hand, the study design without an active control group raises the likelihood of confounding positive with placebo effects. Chess has also been used to treat attention deficit hyperactivity disorder (ADHD). In the light of the positive outcomes, it was argued that playing chess might be of assistance in the treatment of this disorder because it might attenuate its symptoms, particularly at high intelligence levels (Blasco-Fontecilla et al., 2016). Apart from a reduced sample size, however, another key limitation of the study was that the attenuation of symptoms was assessed with parents' self-reports instead of using clinical interviews, which might imply a bias associated with parents' expectations about the effectiveness of the intervention.

Other interventions have been addressed to chess players themselves, particularly with the intention of improving chess performance in high-level competition. For example, acceptance and commitment therapy (ACT), which is a kind of therapeutic intervention aimed at experiential avoidance disorders, has been shown to improve the chess performance of a group of promising young chess players involved in high-level competition (Ruiz & Luciano, 2009). Seven young chess players showed a meaningful improvement in their performance as measured by the Elo rating after the ACT intervention conducted with them. These findings suggest in addition that ACT-based interventions might be a successful alternative to the cognitive-based interventions predominating in current sport psychology.

The usefulness of chess has also been emphasized for teaching certain desirable skills for students with disabilities. Concentration, problem identification and problem solving, planning, creativity, and lucid thinking are important requirements for school daily life. It is likely, however, that these sorts of skills may be very difficult to teach through conventional instructional methods to students with disabilities (Storey, 2000). There are indeed some experiences with chess focused on children with special education needs that are worthy of mention. For example, the study from South Korea describes the implementation of a chess instruction intervention with children at risk of academic failure and with learning disabilities (Hong & Bart, 2007). The research design comprised a random assignment of students to either a chess instruction intervention with one ninety-minute session a week for three months (experimental condition, $n = 18$), or to regular school activities after class (control condition, $n = 20$). The findings of the study indicate that there were not significant differences in the outcomes from the Raven's Progressive Matrices test, or in a test of non-verbal intelligence. These null findings are partly attributed to the limited time of the chess intervention and the rather low level of chess skill of the students in the experimental condition. In any case, the findings fail to support the beneficial effects of chess instruction for children at risk of academic failure and with learning disabilities.

More positive findings are reported in a chess instruction intervention in Germany lasting a complete academic year. This intervention aimed to improve calculation and concentration abilities in a group of students with learning disabilities and who had IQ scores in the low range of 70 to 85 (Scholz et al., 2008). One weekly hour of chess lessons was delivered to an experimental group ($n = 31$), whereas another control group ($n = 22$) followed regular academic lessons. Improvements were observed in calculation abilities such as simple addition and counting, while no improvements were observed in concentration abilities. This study concludes that skills derived from chess lessons can be transferred to improve the basic mathematical skills of children with learning disabilities.

Another chess instruction intervention in the United States that lasted thirty weeks evaluated the mathematics achievement of a group of thirty children with health, auditory, learning, emotional/behavioural, speech, or autism impairments, who were regularly receiving special education services (Barrett & Fish, 2011). These students were assigned to either comparison or treatment groups, whereby the comparison group received the general mathematics curriculum, and the treatment group received chess lessons focused on teaching the rules of the game, together with strategic and tactical concepts connected to mathematics contents. The findings suggest that, although it was exposed to a lesser extent to the general mathematics curriculum, the treatment group was not outperformed by the comparison group in any of the measures of mathematical achievement. Conversely, the treatment group outperformed the comparison group in the final grades in mathematics, in an overall scale score of the maths achievement test, and in two additional subtests involving other mathematical contents such as numbers, operations, quantitative reasoning, and probability and statistics. These findings are considered to support the notion of some sort of transfer from chess skills to mathematical skills for the treatment group.

A more recent study from Turkey explored the impact of chess instruction for visually impaired children in secondary education on mathematics achievement (Aydin, 2015). The chess instruction intervention extended throughout twelve weeks, with one four-hour session being delivered to a group of fourteen children every week, and compared to another group of twelve children who did not follow the chess instruction. The outcomes were analysed by comparing both groups, evaluating whether there were meaningful changes between the autumn and spring academic semesters. There was a meaningful improvement in mathematics achievement in the chess group across both semesters, whereas the chess group obtained a higher mathematics achievement than the non-chess group in the spring semester only. Thus, the findings are regarded as supporting the view that the chess intervention contributed to improving mathematical achievement.

These latter studies could well be considered also in the next section about education and school. It should be highlighted, however, that the groups of children participating in these interventions underwent acute disorders that impaired their vital development meaningfully. Despite that, they were enrolled in regular special education schooling. There have also been plenty of experiments using chess instruction to help the academic schooling of children without these sorts of disorders. This body of evidence is addressed within the next section.

## 10.3 Education and School

It is argued that chess impinges on a range of desirable attributes on the part of schoolchildren regarding their academic, cognitive, and social development. The benefits of chess have been highlighted particularly at early school stages. Back in the early 1980s a study with forty students randomly assigned to chess teaching or non-chess teaching reported that the students involved in chess had significantly better academic results after the chess instruction compared with the students uninvolved in chess (Christiaen & Verhofstadt-Denève, 1981). Throughout the past three decades there have been several similar chess instruction interventions aimed at determining whether they exert tangible effects on the performance and development of schoolchildren. For example, some doctoral dissertations have addressed a variety of topics relating to chess and education in schools, such as the transfer of skills gained in chess to academic achievement (Rifner, 1992), the differential influence of chess instruction depending on economic status (Eberhard, 2003), the design of instructional materials related to academic subjects (Fernández-Amigo, 2008; García & Blanch, 2016), or the impact of chess club participation on scholastic achievement (Garcia, 2008).

There have also been compelling large-scale initiatives to teach chess to schoolchildren in order to improve academic results, cognitive abilities, and social adjustment factors. Most of these programmes have been designed to work with chess contents essentially addressed to primary education. For example, a study funded by the Scottish Executive Education Department and Aberdeen City Council implemented a three-year programme centred on chess coaching. This programme aimed at assessing the relationship between children's general learning, reading and language, thinking skills, social adjustment, parental support for study at home, and chess tuition (Forrest et al., 2005). Three groups of eighteen students each were compared after being exposed to different interventions: (1) chess coaching; (2) computer games; and (3) no chess or computer games. The findings, compared in several measures at the beginning and at the end of the study period, indicate positive changes in reading and arithmetic skills, and in social adjustment in the chess coaching group. There were, in addition, significant improvements for the computer games group in vocabulary and arithmetic skills.

The after-school programme Chess for Success (CFS) in the United States is another project that provides chess instruction to elementary and middle-grade students who wish to learn how to play chess (Yap, 2006). The project aims explicitly to improve academic achievement, self-esteem, and skills related to patience and analytical problem solving. This intervention lasted fifty hours and involved a large number of students ($n$ = 321) from several schools in Oregon in 2004, 2005, and 2006, with an experimental group ($n$ = 233) exposed to the chess intervention, and a control group ($n$ = 88). The findings suggest more robust significant improvements in mathematics than in reading across the three-year period for the experimental group than for the control group. Moreover, the CFS programme was considered to be an excellent experience for stimulating students' involvement in schooling and the participation of a large amount of girls in the project. This report also highlights the need to evaluate the effects of the programme in the long term, however.

The Chess Challenge Program is an initiative to implement chess instructional activities sponsored by the After School Activities Partnerships, based in Philadelphia (Pennsylvania). A report from this institution describes the evaluation of a chess intervention during the 2007/8 and 2008/9 academic years that was implemented in seventeen schools (DuCette, 2009). The objective of this specific assessment was to determine the extent to which participation in the project improved academic achievement and school-related behaviour. The participants in the project were compared against an equivalent group of students matched in school, grade, gender, and ethnicity. The findings show that students participating in chess had higher scores in mathematics and reading across the two assessment years, better school-related behaviour ratings, and a lower number of absences from school. In addition, these findings demonstrate the effectiveness of the chess programme in improving these academic and behavioural areas.

The largest study not just in terms of the amount of students involved but also regarding the comprehensiveness of its evaluation was that carried out by the United Kingdom's Education Endowment Foundation (EEF). The EEF was established in 2011 as a grant-making charity aiming to close the gap between family income and educational outcomes. The EEF conducted a large-scale chess intervention in 100 schools with over 4,000 students and a comprehensive evaluation of a randomized controlled trial design (Jerrim et al., 2016). The main finding from this study is that chess instruction has little lasting influence on children's academic achievement. The evidence obtained was unsupportive of a positive impact on mathematics, reading, or science achievement one year after the intervention. Furthermore, the findings from the evaluation of this chess intervention underline two additional key points. First, chess tutors should be competent in class management and teaching in order to successfully engage children in chess instruction. Second, future studies should focus on the purported benefits of chess to fields other than academic achievement, with a particular emphasis on children's well-being and non-ability skills.

Table 10.1 summarizes several of these studies ($n = 19$) selected on the basis of three main criteria. First, the studies had to describe a chess instruction intervention with school-aged children, aimed to improve academic achievement, cognitive skills, or socio-affective competences. Second, only studies conducted beyond the year 2000 were included. Third, the study had to be published either in a specialized journal, or as a working report easily accessible through the internet.

For each study there is the age and number of participants, the academic, cognitive, or social target areas, the duration of the chess intervention in hours, the male to female ratio, and the country of origin. The last column indicates in addition whether the reported effect of the chess instruction intervention on the target areas was either positive or null. All the interventions were undertaken with children from five to sixteen years old, with sample sizes being highly variable and ranging from only twenty-six participants to a sizeable 4,009 sample. Although several studies had very small sample sizes, there were some notable exceptions with very large samples for instructional intervention studies of this kind, such as two studies from Italy (Sala et al., 2015; Trinchero & Sala, 2016), and the large randomized control trial from England (Jerrim et al., 2016).

Most of the interventions focused on the academic subject of mathematics, probably because several mathematical skills are thought to match chess skills. Two studies addressed several areas with an extensive sample (Aciego et al., 2012, 2016). One study with very young students, from five to six years of age, addressed in turn spatial concepts (Sigirtmac, 2012). The duration of the chess interventions ranged from ten to ninety-six hours. The male to female ratio was more balanced than the chess studies about other topics reviewed so far. The studies from Iran and Canada report male samples only (Kazemi et al., 2012; Voyer et al., 2018), whereas most of the remaining studies report at most

Table 10.1 *Overview of studies describing a chess instructional intervention for schoolchildren (M:F = male to female ratio)*

| Study | Age sample* | N | Target area | Hours | M:F | Country | Effect |
|---|---|---|---|---|---|---|---|
| Aciego et al., 2012 | 6–16 | 210 | Social/emotional enrichment | 96 | – | Spain | + |
| Aciego et al., 2016 | 6–16 | 230 | Social/emotional enrichment | 96 | – | Spain | + |
| Aydin, 2015 | – | 26 | Mathematics | 48 | 2:1 | Turkey | + |
| Barrett & Fish, 2011 | 8–12 | 31 | Mathematics | 30 | 2:1 | United States | + |
| DuCette, 2009 | 6–15 | 151 | Mathematics, reading, behaviour | – | 2:1 | United States | + |
| Forrest et al., 2005 | M = 8.8 (–) | 54 | Reading, arithmetic, social adjustment | 37 | – | Scotland | + |
| Gliga & Flesner, 2014 | M = 9.7 (0.77) | 38 | Mathematics, language | 10 | 1:1 | Romania | + |
| Hong & Bart, 2007 | M = 9.7 (–) | 38 | Mathematics, reading, writing | 20 | 2:1 | South Korea | Null |
| Jerrim et al., 2016 | 9–10 | 4,009 | Mathematics, science, reading | 30 | 1:1 | England | Null |
| Joseph et al., 2016 | – | 100 | Mathematics, language, science, social | – | – | India | + |
| Kazemi et al., 2012 | – | 180 | Meta-cognitive abilities, mathematics | 96 | 180:0 | Iran | + |
| Rosholm, Mikkelsen, & Gumede, 2017 | M = 10 (–) | 482 | Mathematics | – | – | Denmark | + |

| | | | | | | | |
|---|---|---|---|---|---|---|---|
| Sala, Gorini, & Pravettoni, 2015 | M = 9 (1) | 560 | Mathematics | 18 | 1:1 | Italy | + |
| Sala & Gobet, 2017 | M = 9 (1) | 233 | Mathematics | 25 | – | Italy | Null |
| Scholz et al., 2008 | 10–11 | 53 | Mathematics, concentration | 24 | 2:1 | Germany | + |
| Sigirtmac, 2012 | 5–6 | 100 | Spatial concepts | 50 | 1:1 | Turkey | + |
| Trinchero & Sala, 2016 | M = 9 (1) | 931 | Mathematics | 19 | 1:1 | Italy | + |
| Voyer, et al., 2018 | 8–9 | 185 | Mathematics | 60 | 185:0 | Canada | + |
| Yap, 2006 | – | 321 | Mathematics, reading, self-esteem, behaviour | 50 | 2:1 | United States | + |

*Note:* * = age range or mean and standard deviation (M(*Sd*)).

a 2:1 male to female ratio. Concerning the geographical area, there are three studies from the United States and Italy, two from Spain and Turkey, and one from Scotland, Romania, South Korea, England, India, Iran, Denmark, Germany, and Canada.

The most striking finding is that only three studies report a null effect from the chess instruction process on the selected target areas (Hong & Bart, 2007; Jerrim et al., 2016; Sala & Gobet, 2017). In contrast, the rest of the studies report a positive effect from the respective intervention in the specific target areas. This outcome represents 84% of the reported interventions, which on the face of it appears to provide overwhelming evidence in favour of a positive influence of chess instruction on the respective target areas. Nevertheless, several of these studies might include fundamental flaws with regard to their methodological design, being far from what has been termed the ideal experiment, diluting some of their main conclusions.

In a critical review of chess instruction and its potential benefits for education, Gobet and Campitelli describe the specific requirements of an ideal experiment for evaluating the effects of chess instruction on education (Gobet & Campitelli, 2006). These requirements comprise a random assignment of participants to experimental groups, a pre-test and a post-test, a placebo group and a do-nothing group, different persons being assigned to the roles of instructor and evaluator, and subjects' blindness to the purposes of the instructional intervention. After reviewing seven studies that had implemented some kind of chess instruction intervention, these authors conclude that all of them failed to abide by these ideal experiment requirements, and, in turn, featured several methodological weaknesses. It should be acknowledged, however, that these requirements are often unrealistic, not only in chess instruction studies but also in general educational and psychological research. In addition, this review highlights several characteristics that might eventually hamper the robustness of the reported findings within this specific field of research. The lack of sufficient details about methods and results, the low quality of studies that prevented publication in peer-reviewed journals, the bias towards emphasizing supportive results only, and the absence of a sound theoretical basis guiding the particular research effort were all crucial missing characteristics to comply with the minimal acceptable scientific research standards. This in-depth critical review concludes by highlighting three main ideas.

1. The effects of chess instruction on other domains of interest remain an open question.
2. Compulsory chess instruction should be disregarded because it might convey motivational problems.
3. Chess benefits to other domains might decrease remarkably with progressive improvements in chess skill.

These findings have also been corroborated in a subsequent critical review of this controversial topic, focusing on the effects of chess on scholastic achievement (Bart, 2014). In accordance with this latter review, however, the basic sequence of chess playing – i.e., position comprehension, pattern induction, evaluation and move selection – is suggested to be likely to involve a transfer potential to learning in mathematics and related fields.

A further quantitative meta-analytical review with most of the aforementioned studies that reported positive effects (see Table 10.1) reports only a weak positive influence of chess instruction on academic achievement and cognitive abilities (Sala & Gobet, 2016). Moreover, it is confirmed again that none of the studies included in the meta-analysis contemplated the requirements derived from the ideal experiment. For example, Sala and Gobet emphasize research designs including a do-nothing group and an active group to control for potential placebo effects. For instance, a more recent work reporting two studies applies such a research design with three different groups (Sala & Gobet, 2017). In the first study, there was an experimental group involved in chess lessons ($n$ = 53), an active control group involved in checkers lessons ($n$ = 83), and a do-nothing group ($n$ = 82) involved in regular school activities. In the second study, participants were randomly assigned to either a chess instruction group ($n$ = 15), a Go game group ($n$ = 15), or a control group ($n$ = 22). In general, these findings are negative. Both studies point out that the hypothetical benefits of chess instruction on mathematical skills are untenable.

Furthermore, alternative research designs for comparing two or more groups of students involved in chess playing have also yielded unsupportive findings regarding the influence of chess on academic achievement. A research work in Australia applied a sophisticated data analysis based on item response theory and hierarchical linear modelling (Thompson, 2003). This study evaluated the hypothesis of whether students playing chess on a regular basis would perform better in the Australian Schools Science Competition, a multiple-choice test designed to assess scientific thinking. While controlling for grade level and an undetermined IQ measure, the findings of this study are unsupportive of this hypothesis, suggesting instead that students who performed better in scientific thinking also scored higher in the IQ measure. Besides, it is argued that those students would have been eventually more interested in chess. Overall, 55% of the variability in the specific scientific thinking test was explained only by individual differences in the grade level and the IQ.

There are two main additional issues to bear in mind regarding the allegedly beneficial influence of chess on academic achievement or on learning other abilities. Both of them should be taken into account regarding chess instructional interventions addressed to regular students and students with special educational needs. The first issue implies the idea of transfer; more specifically, whether the abilities learned in chess are readily transferable to different academic and cognitive fields. The second issue implies the idea of statistical power in comparative studies; more specifically, whether a considerable proportion of the studies supporting the view that there are beneficial effects from chess interventions are likely to be statistically underpowered. These two issues are addressed in greater depth in the next two sections.

## 10.4 Transfer

Transfer in a learning environment implies that the changes in an ability to perform a task influence the ability to perform the same task under different conditions or the ability to perform a different task. A transfer model comprising two measures of performance ($x$, $y$) and practice ($p_x$, $p_y$) in two tasks can be defined with a function such as $y = \varphi\ (x, p_x, p_y)$, indicating that performance in the task $y$ is a function of the performance in the task $x$, and the amount of practice in the two tasks. Different transfer functions are possible by varying the implied parameters. In addition, two main hypotheses have been suggested to account for the nature of transfer learning from one task ($x$) to another task ($y$). First, the most important factors involved in transfer effects are cognitive abilities. Second, different abilities enforce different effects at different stages of learning (Ferguson, 1954, 1956). In this view, general intelligence correlates with positive transfer. Individuals scoring higher in cognitive abilities tests are more likely to benefit from practice and improve their performance on different but related tasks.

Furthermore, it has been found that aptitude by treatment interactions arise in learning situations bearing potential transfer functions with both simple tasks (Skanes et al., 1974) and complex tasks (Espejo, Day, & Scott, 2005). The aptitude by treatment interaction (ATI) consists of the differential predictions arising in the cognitive aptitude for learning from instruction. This theory has exerted a considerable impact within the field of instructional psychology, emphasizing the crucial role of individual differences in instructional processes (Snow & Lohman, 1984; Snow & Swanson, 1992). The concept of transfer is a central tenet of this theory. Most studies describing chess instructional experiences tend to overlook crucial ATI components and assumptions, however. For example, within-class and between-class differences are likely to influence aptitude by treatment interactions, even though evaluating these sorts of effects has been generally ignored within the chess instruction literature (Cronbach & Webb, 1975; Gustafsson, 1978). A remarkable exception, however, is that described in a doctoral dissertation about transfer, where it is reported that transferring what was learned in chess to poetic analysis depended on individual differences in cognitive ability (Rifner, 1992). This doctoral dissertation emphasizes that transfer occurred to a greater extent for students of above-average ability.

There are in addition three instructional implications that should be considered in chess instruction experiences to enhance the likelihood of transfer, adapted to the peculiarities and circumstances of every particular situation. First, instructional treatments should vary in their 'incompleteness'. Learners with lower scores in cognitive abilities would benefit more from explicit, direct, and detailed structured instruction. Learners with higher scores in cognitive abilities would benefit more from incomplete and unstructured instruction. Second, training for learners with lower scores in cognitive abilities should be aimed at developing very specific learning skills, together with flexible adaptation to other learning problems. Third, the evaluation of chess instruction should follow up the development in each treatment, advise when learners with low scores in cognitive abilities should be promoted to incomplete and unstructured instruction, and assess the influence of individual differences in motivational and affective aspects of the instructional intervention.

From a cognitive point of view, there seem to be three main mechanisms of knowledge transfer – analogy, knowledge compilation, and constrain violation – which can be readily amenable to the characteristics of the chess domain (Nokes, 2009). Concerning transfer from learning chess to performance in other tasks in academic and cognitive domains, seven types of transfer effects have been put forward (Scholz et al., 2008): positive, negative, lateral, vertical/hierarchical, sequential, specific, and unspecific. For example, recent studies have argued that, because chess playing involves cognitive skills dealing with focusing, visualizing, thinking ahead, analysing abstract options, and planning, chess should influence the academic achievement of students without exception (Joseph, Easvaradoss, & Solomon, 2016). Furthermore, a study with a large sample from Denmark ($n = 482$) reports that substituting a weekly lecture of mathematics with chess instruction contributed to significant increments in mathematics achievement, and explicitly highlights that the transfer of skills and knowledge across both domains was indeed possible (Rosholm et al., 2017). Nevertheless, this effect is also argued to be constrained to students who were bored and unhappy with traditional curricular contents.

Putting aside the controversy of whether chess instruction interventions addressed to schoolchildren improve academic achievement, cognitive abilities, and socio-affective skills, there is a crucial unknown in this field of research. Whether skills acquired in chess are transferable to other domains of intellectual activity remains unclear and is an unsettled question. Empirical evidence points against the eventual transfer from chess to other domains. For example, there are findings from a large sample of chess players ($n = 6{,}112$) suggesting that chess skills are greatly constrained and specific to the game. These findings from a game theory experiment (beauty contest) suggest that chess-playing skills are not associated with a particularly high level of rational choice (Bühren & Frank, 2012). Another study within the game theory approach emphasizes that there were no differences in performance in the race to 100 game among players with different levels of chess skill, from grandmasters to players below 2000 Elo points (Levitt et al., 2011). Similarly, a two-experiment study comparing chess and non-chess players in a planning performance task such as the Tower of London does not lend support to the view of transfer from planning abilities in chess to planning abilities into other domains (Unterrainer et al., 2011). Moreover, empirical findings about chess practice suggest a potential generalization to other intellectually demanding domains, even though the transfer of skills learned in chess to learning skills in other domains is seriously questioned. Besides, studying chess to improve performance in other domains was firmly discouraged (Campitelli & Gobet, 2008). One important hindrance impairing transfer could be, for instance, the extensive amount of time required to learn chess skills, which would prevent the allocation of additional time to learning skills in other domains.

From this point of view, transfer across domains would be feasible only with very similar cognitive environments in the implied domains, and would be unlikely with very dissimilar cognitive environments in the implied domains. The quantitative meta-analysis about the eventual transfer of chess instruction to academic and cognitive skills analysed the reported findings of twenty-four studies (Sala & Gobet, 2016). The main findings indicate a rather unconvincingly moderate effect size, with a stronger effect on mathematics than on reading, and a positive effect regarding the duration of the instructional intervention. Nevertheless, and as with earlier reports (Bart, 2014; Gobet & Campitelli, 2006), it is claimed that none of the analysed studies based their conclusions on an ideal research design. Most of these research works rely on the comparison of either two groups (chess instruction and control) or two situations (pre-test and post-test), with some sort of hypothesis contrast. Therefore, another methodological requirement that might additionally undermine the conclusions in this field of research is that related to statistical power.

## 10.5 Statistical Power

The statistical power of a hypothesis contrast test ranges from 0 to 1. This value indicates the probability of rejecting the null hypothesis by the test when the alternative hypothesis is true: Pr (reject H0 | H1 is true). Put in other words, power is the probability of correctly rejecting the null hypothesis when it actually should be rejected because the alternative hypothesis is a better explanation of the observed data. Statistical power is typically represented as $1 - \beta$, where $\beta$ is termed the type II error, representing the probability of failure to reject the null hypothesis when it is false (Hair et al., 2010). Statistical power is determined by three factors: effect size, alpha, and sample size. The effect size is a quantitative value that describes the magnitude of the studied phenomenon. Usually the effect size is described in standardized units, such as a value of 0.5, which indicates an effect size of one-half of a standard deviation. For example, Jacob Cohen's *d* measure gives an idea of the magnitude of the effect size for a difference between means, and it is computed with the difference of the means divided by the standard deviation of the pooled data (Cohen, 1988). The second determinant factor of statistical power is the alpha value. The alpha is also termed the type I error, and it represents the probability of rejecting the null hypothesis when it is true: Pr (reject H0 | H0 is true). Type I and type II errors are inversely related. Reducing type I error increases type II error, and, consequently, decreases the power of the test to detect a meaningful effect if it exists. In psychology and education, alpha levels of 0.05 or 0.01 are the most usual choices. The sample size is the third intervening factor in statistical power. Increasing sample size at any level of alpha contributes to increased statistical power. The growing statistical power of the test results in smaller effects becoming incrementally significant, however. Thus, smaller sample sizes tend to be insensitive to statistical tests, whereas larger sample sizes tend to be oversensitive.

The studies summarized in Table 10.1 were submitted to post-hoc analyses, aiming to estimate the available statistical power in the main comparisons performed in these studies. The analyses were implemented with the pwr package from the R computer software (R Development Core Team, 2015). Specifically, the power.t.test() function was used, which includes sample size, effect size, variance, significance, and power parameters. The analyses were performed by fixing four of these parameters: the sample size, the effect size, the variance, and the significance. More specifically, statistical power was estimated by combining two effect sizes, three variances, and one significance level. The sample sizes were fixed as reported in the specific comparison conducted in each study: either experimental versus control groups, or pre-test versus post-test. The two effect sizes (*d*) for a comparison of the means were fixed at 0.5, which represents a medium effect, and at 0.8, which represents a large effect (Cohen, 1988). The three variances were fixed at one, two, and three standard deviations, which comprises the ranges of most measures employed in the analysed studies. The significance level was fixed at 0.05, which is also the most commonly accepted in educational and psychological research. Hence, for each study with two sample sizes, corresponding to the experimental and control groups, the statistical power needed to detect a medium or a large effect ($d = 0.5$ or $d = 0.8$) was estimated at one, two, and three standard deviations.

The findings of the statistical power analyses are shown in Table 10.2. The first two columns indicate each study with its respective sample sizes (*N*). The next three columns indicate the required statistical power to detect a medium effect size ($d = 0.5$) at one, two, and three standard deviations. The next three columns, on the right, indicate the required statistical power to detect a large effect size ($d = 0.8$) at one, two, and three standard deviations. The entries in Table 10.2 indicate that the statistical power to detect a medium effect size is

Table 10.2 *Post-hoc analyses of statistical power (1 – β) of studies describing a chess instructional intervention for schoolchildren (*d *= 0.5 and 0.8, α = 0.05). The studies with only one sample size conducted comparisons of experimental and control groups of equal sample sizes*

| | | *d* = 0.5 medium | | | *d* = 0.8 large | | |
|---|---|---|---|---|---|---|---|
| Study | N | Sd = 1 | Sd = 2 | Sd = 3 | Sd = 1 | Sd = 2 | Sd = 3 |
| Aciego et al., | 170 | 1.0 | 0.63 | 0.33 | 1.0 | 0.96 | 0.69 |
| 2012 | 60 | 0.76 | 0.27 | 0.15 | 0.99 | 0.58 | 0.30 |
| Aciego et al., | 110 | 0.96 | 0.45 | 0.23 | 0.99 | 0.84 | 0.50 |
| 2016 | 60 | 0.78 | 0.27 | 0.15 | 0.99 | 0.58 | 0.30 |
| Aydin, 2015 | 14 | 0.25 | 0.09 | 0.06 | 0.53 | 0.17 | 0.10 |
| | 12 | 0.22 | 0.08 | 0.06 | 0.47 | 0.15 | 0.09 |
| Barrett & Fish, | 16 | 0.28 | 0.10 | 0.07 | 0.59 | 0.19 | 0.11 |
| 2011 | 15 | 0.26 | 0.10 | 0.06 | 0.56 | 0.18 | 0.10 |
| DuCette, 2009 | 151 | 0.99 | 0.58 | 0.30 | 0.99 | 0.93 | 0.64 |
| Forrest et al., 2005 | 18 | 0.31 | 0.11 | 0.07 | 0.64 | 0.21 | 0.12 |
| Gliga & | 20 | 0.34 | 0.12 | 0.07 | 0.69 | 0.23 | 0.13 |
| Flesner, 2014 | 18 | 0.31 | 0.11 | 0.07 | 0.64 | 0.21 | 0.12 |
| Hong & Bart, | 20 | 0.34 | 0.12 | 0.07 | 0.69 | 0.23 | 0.13 |
| 2007 | 18 | 0.31 | 0.11 | 0.07 | 0.64 | 0.21 | 0.12 |
| Jerrim et al., | 1,965 | 1.0 | 1.0 | 1.0 | 1.0 | 1.0 | 1.0 |
| 2016 | 1,900 | 1.0 | 1.0 | 1.0 | 1.0 | 1.0 | 1.0 |
| Joseph et al., | 48 | 0.68 | 0.23 | 0.12 | 0.97 | 0.49 | 0.25 |
| 2016 | 52 | 0.71 | 0.24 | 0.13 | 0.98 | 0.52 | 0.27 |
| Kazemi et al., | 86 | 0.90 | 0.37 | 0.19 | 0.99 | 0.74 | 0.41 |
| 2012 | 94 | 0.93 | 0.40 | 0.21 | 0.99 | 0.78 | 0.44 |
| Rosholm et al., | 323 | 1.0 | 0.89 | 0.56 | 1.0 | 0.99 | 0.92 |
| 2017 | 159 | 1.0 | 0.60 | 0.31 | 1.0 | 0.94 | 0.66 |
| Sala et al., 2015 | 309 | 1.0 | 0.87 | 0.54 | 1.0 | 0.99 | 0.91 |
| | 251 | 1.0 | 0.80 | 0.46 | 1.0 | 0.99 | 0.85 |
| Sala & Gobet, | 53 | 0.72 | 0.25 | 0.13 | 0.98 | 0.53 | 0.27 |
| 2017 | 82 | 0.89 | 0.36 | 0.18 | 0.99 | 0.72 | 0.40 |
| | 98 | 0.94 | 0.41 | 0.21 | 0.99 | 0.80 | 0.46 |
| Scholz et al., | 31 | 0.49 | 0.16 | 0.09 | 0.87 | 0.34 | 0.18 |
| 2008 | 22 | 0.37 | 0.13 | 0.08 | 0.74 | 0.25 | 0.14 |
| Sigirtmac, 2012 | 50 | 0.70 | 0.24 | 0.13 | 0.98 | 0.51 | 0.26 |
| Trinchero & | 320 | 1.0 | 0.88 | 0.56 | 1.0 | 0.99 | 0.92 |
| Sala, 2016 | 220 | 1.0 | 0.74 | 0.41 | 1.0 | 0.99 | 0.80 |
| | 391 | 1.0 | 0.94 | 0.64 | 1.0 | 0.99 | 0.96 |

Table 10.2 *Cont.*

| | | *d* = 0.5 medium | | | *d* = 0.8 large | | |
|---|---|---|---|---|---|---|---|
| Study | N | Sd = 1 | Sd = 2 | Sd = 3 | Sd = 1 | Sd = 2 | Sd = 3 |
| Voyer et al., | 70 | 0.84 | 0.31 | 0.16 | 0.99 | 0.65 | 0.35 |
| 2018 | 73 | 0.85 | 0.32 | 0.17 | 0.99 | 0.67 | 0.36 |
| Yap, 2006 | 233 | 1.0 | 0.77 | 0.43 | 1.0 | 0.99 | 0.82 |
| | 88 | 0.91 | 0.38 | 0.19 | 0.99 | 0.75 | 0.42 |

lower than that to detect a large effect size, and is also lower at higher variabilities. Thus, the lower values for the statistical power to detect a medium effect size are found at three standard deviations, whereby only one study reaches a perfect statistical power (Jerrim et al., 2016). At the medium effect size with two standard deviations, there is only one additional study with a statistical power equal to or above 0.8 in all the participating samples (Sala et al., 2015), while at one standard deviation only about half the studies reach a statistical power equal to or above 0.8 in all samples. The values for the statistical power to detect a large effect size are somewhat higher. In the most favourable condition, with one standard deviation, there are thirteen studies with a statistical power equal to or above 0.8, representing 68% of the total pool of studies. In the most unfavourable conditions, with two and three standard deviations, there are only five and three studies meeting the 0.8 statistical power criterion, representing 26% and 16% of the total pool of studies, respectively.

Overall, these findings indicate that the studies addressing the effects of chess instruction interventions to improve academic results, cognitive abilities, and other desirable attributes for the development of schoolchildren are generally underpowered from the statistical point of view. These analyses suggest that claiming to reject the null hypothesis with a low statistical power seriously undermines the validity of this decision. Within the studies evaluating the impact of chess on the aforementioned outcomes, only the large intervention in the United Kingdom reports the available statistical power in advance (Jerrim et al., 2016). Furthermore, and due to its large sample size, although this study has the largest statistical power throughout all the analysed conditions, it could be claimed that it might impose an oversensitive test on the analysed relationship, even though it is one of the few studies reporting null effects of the chess intervention in the long term. The chess intervention was carried out between October 2013 and July 2014, whereas the evaluation of the effects was carried out in June 2015. Another important limitation of this group of studies relates to those studies that used different sample sizes for their experimental and control groups. In general, experimental groups had larger sample sizes than control groups. Therefore, the reported positive effects of chess instruction might be biased towards experimental groups because of their typically having a higher statistical power than control groups. Thus, it is advisable to consider very similar sample sizes in research designs comparing chess instructional interventions that include experimental, control, and placebo groups.

Chess instruction for schoolchildren has become very popular in the past ten years. The abilities learned in chess are supposed to have beneficial effects in several areas that are important for child development, such as cognitive abilities, school and social adjustment, concentration, self-control, and many other desirable outcomes with which most parents and schoolteachers would be the happiest people in the world. In general, the supporters of these sorts of initiatives are especially passionate and vehement when they are advocating for the salutary effects of chess on children. The supporting evidence for these effects in terms of scientific data is as weak as that offered for brain-training programmes, however (Chabris, 2017). Attention should be paid, therefore, to four main considerations when implementing chess instruction intended to improve learning in areas other than chess. First, the intervention should be guided by a sound theoretical basis, such as the aptitude by treatment interaction approach. Second, the research design should be planned according to ideal experimental premises. Third, transfer issues should be defined in advance, while evaluating the potential transfer effects at several points in time before, and after, the intervention. Fourth, statistical power should be considered prior to the implementation of the intervention, by describing the expected effect sizes, samples, and variability in the target participants.

# 11

# Concluding Remarks

While I am writing these lines, the 2018 World Chess Championship is in progress. Magnus Carlsen, with an Elo rating of 2835 points, is confronting Fabiano Caruana, with an Elo rating of 2832 points. Millions of people, through the internet, are actually following this unique match. Moreover, the event is being extensively covered in specialized internet portals that offer a massive but well-organized amount of information about chess – from the outcomes of the latest chess contests, to courses on chess tactics and positional play, or the most cunning novelty in the Dragon variation from the Sicilian opening played in the more recent stronger chess competitions. Most appealing, however, is the opportunity to play all sorts of game modalities, fast and slow, with a wide array of competitors from all over the world. Back at the beginning of the twentieth century, correspondence chess was for some people a charming modality: bizarre, very slow chess playing. A correspondence chess game could last for months, or even years (Harding, 2009). Things have moved rapidly. Research into chess psychology has also experienced a relatively quick evolution. Since the early works from Binet, an influential precursor of intelligence testing, several works are at present looking at what is happening inside the brains of chess players with sophisticated techniques.

Chess is the prototypical intellectual game. It has been considered the *Drosophila* for cognitive psychology and expertise research, and it has also been informative for artificial intelligence. The chess environment is highly appealing for psychological research because of its systematic and universally accepted system of rules, which enhances its ecological validity. Moreover, a valid and reliable rating of chess strength is relatively easy to estimate, the Elo rating, which is systematically recorded from a large number of chess players around the world. There is in addition a wealth of nationwide available data about chess skill from large groups of active chess players extending over relatively long periods. Furthermore, major statistical developments such as big data approaches permit the examination of huge amounts of chess data, from either chess-playing skill or actual chess games (Vaci & Bilalić, 2017). Moreover, information technologies such as the expansion and generalization of the internet have stimulated greater and more fluid communication between researchers, and the development of data and experiment-sharing policies and initiatives such as the open science framework. As a consequence, behavioural research using chess as a model now has a relatively long history, extending over more than a century.

This book compiles and summarizes this body of research in a single volume. Unlike other recent books about the scientific study of the psychology of chess (Bilalić, 2017; Gobet, 2019), the current book emphasizes individual differences along two main lines of enquiry. On the one hand, the book describes the body of research that has employed central constructs from differential psychology such as intelligence and personality. On the other hand, the book underlines the main findings that have been informative concerning common topics of interest for differential psychology, such as psychophysiology and brain functioning, heredity and environmental determinants of behaviour, ageing, and sex differences. Furthermore, the book describes several works that have used chess to address problems in the business, health, and educational fields.

There are several issues that should be particularly underlined in the light of the extant body of research about individual differences in chess. Some of them might be considered to be well-established facts. Other issues could be taken as limitations that, as far as is known, have also been remarked upon in several of the studies reviewed here. Yet other issues are unknowns that have been identified along the way. Rather than placing these remarks in a particular order, they are exposed in accordance with the topics addressed throughout this book – a more logical sequence, just like the solution to the chess problem proposed in the introductory section (see Figure 1.1). For some of these issues, whether it is a known and well-established fact, a research limitation, or an unknown may be obvious. For other issues it may not. These latter ones are left for you, the reader, to decide.

1. Individual differences are evident in daily life in most areas involving human behaviour. Chess is no exception. Because chess is such a well-defined domain with universal regulations, and entailing the interplay of several individual differences in behaviour, it is an optimal model for analysing individual differences in action, free of the constraints imposed by artificially and unrealistic laboratory conditions.
2. The Elo rating is a powerful and ecologically reliable and valid indicator of intellectual performance. There are plenty of archival data of the Elo rating from the FIDE, and from several other chess federations in several countries. These databases are a rich source of information about individual differences in intellectual performance, which could and should be used to address crucial themes and hypotheses within this field of research.
3. Alternatively, there are more efficient quantitative rating systems available at present. For example, the Universal Rating System (URS) appears as an interesting alternative to the Elo rating system, because it contemplates the

outcomes in chess games at different time controls, fast and slow, to calculate a single rating value for each individual chess player.

4. Apart from the availability of chess skill data, the availability of a vast amount of chess games also offers an attractive window for looking at the products of intensive intellectual work. For instance, time controls allow us to examine behaviour in decision-making when constraining the availability of thinking time. Moreover, different chess openings might underlie different playing styles, which are probably associated, in turn, with individual differences in personality. There is also the possibility of comparing the produced chess games and their outcomes in accordance with sex, age, or geographical region.
5. Psychophysiology studies lack sufficient emphasis on relating the findings to individual differences in chess skill as measured by the Elo rating. In addition, future studies with psychophysiological recordings could include psychometric measures of individual differences in intelligence and personality to evaluate potential interactions between both kinds of variables, and to provide a more comprehensive portrait of the hypothesized relationships.
6. There is a remarkable paucity of studies analysing hormonal functioning during chess playing. Hormones are implicated in certain cognitive abilities that have also been related to chess performance. This area of research is largely unexplored, even though it might provide additional useful data for hypothesis testing and theory building.
7. Just as with other experimental studies in psychological research, brain-imaging research works tend to use very small sample sizes, hampering statistical power. Because these studies typically tend to be more expensive, it would be advisable to enhance the collaborative effort of the several researchers working in this field with experiment and data sharing across different laboratories. This may contribute to increasing the sample size of the studies and to improving statistical power. Moreover, as far as is known there are no studies comparing the brain functioning of male and female chess players with modern techniques. This is a notable gap within the brain-imaging field focused on chess playing.
8. Individual differences in both pattern recognition (system 1) and search (system 2) thinking processes underlie individual differences in chess skill. The predominance of one process over the other may depend on the specific circumstances of the game – i.e., opening, middlegame, endgame – but also on the individual differences of the two contenders during a chess game. An interesting and relatively unexplored line of research would be to look at whether individual differences in cognitive abilities, personality, or interests are associated with pattern recognition and search in each of the aforementioned playing circumstances.

9. As with any other intellectual human endeavour, chess skill and chess performance are eventually linked in a complex interplay between several objective variables involving individual differences. These variables comprise cognitive abilities, personality dispositions, interests, motivation, emotional regulation, a precocious starting age, experience, and practice. There is a marked scarcity of studies comparing the magnitude of these relationships, however. An interesting line of future research concerns comprehensive analyses that include the simultaneous evaluation of these traits when examining their interplay in predicting the structure and development of chess skill and chess performance. The intelligence as process, personality, interests, and intelligence as knowledge theory is just such a comprehensive framework for addressing the impact of individual differences in broad constellations of psychological traits on adult intellectual development. Hence, research work within this field should consider the inclusion of measures of traits from as many of the broad domains of the PPIK theory as possible.
10. Individual differences in cognitive abilities appear to exert a stronger influence on chess skill than personality factors. Numerical ability is the most robust cognitive ability associated with chess skill. Nevertheless, the effect is rather weak, and should be further replicated when contrasted with other abilities. Moreover, visuospatial abilities have long been considered important for chess skill. Visuospatial abilities appear to be much more relevant for children than for adults, however. Moreover, motivation and emotional regulation influence chess performance and chess skill meaningfully.
11. There are fewer studies linking traits from the personality domain with chess skill than from the intelligence domain. Furthermore, the evidence supporting the association of personality traits with chess skill is also much weaker. This suggests that individual differences in personality are likely to exert a lower direct effect on chess skill. In contrast, personality factors might act as mediators between cognitive abilities or practice and chess skill. Moreover, individual differences in personality can be particularly influential for chess skill at early stages of the chess career. Future research about the personality of chess players could address the role of personality in chess playing as a mediator and in the initial stages of chess careers.
12. Ideally, studies about individual differences in chess should include larger samples than those typically used in experimental studies. Lower sample sizes tend to hamper statistical power to a great extent. This problem is particularly relevant in psychophysiological and brain-imaging studies. It is usually found, however, that sample sizes in chess studies are below 100 individuals. Again, collaboration between different researchers might

contribute to increasing the sample sizes of chess players, thus improving the statistical power of empirical tests.

13. Apart from practice, expert chess performance depends on natural talent, which comprises the interplay of genetic and environmental factors. Both kinds of factors are highly influential on several individual traits – such as cognitive abilities, personality dispositions, and motivational and emotional factors – that are important for chess performance. Deliberate practice explains a decent amount of the variability in expert chess performance. The view that deliberate practice alone is able to explain the substantial individual differences in expert performance is not supported by a large body of conceptual and empirical evidence, however. Conversely, natural talent is likely to contribute to obtaining higher benefits from deliberate practice activities to a great extent.
14. The remarkable gender gap regarding the participation of men and women in chess is analogous to that found for science, technology, engineering, and mathematics fields and occupations. The smaller amount of women in chess is likely to originate in the interplay between biological factors involving hormones, with individual differences in vocational interests, and cultural differences across countries. Another relatively unexplored issue is whether women chess players may be more likely to place a stronger emphasis on life goals focused in the family realm rather than pursuing a long and demanding chess career, just as is reported about the underrepresentation of women in other STEM fields.
15. Several studies that relate to chess instruction programmes in the school context report meaningful effects deriving from chess in academic achievement and in other areas that are important for child development. In general, however, there are three main problems with these kinds of studies: the research design; the transfer of chess skills to target-area skills; and their typically low statistical power. These studies rely on very simple and limited research designs. The transfer processes of the competences acquired in chess to academic, cognitive, or social areas are unclear, have been extensively criticized, and should be treated with caution before making consistent generalizations. Moreover, the conclusions of most of these studies rely on statistical contrasts that are distinctly underpowered.
16. Chess instruction activities in the schools should abide by the principles of a sound theoretical framework, such as the aptitude by treatment interaction approach. Besides, chess instruction programmes addressed to schoolchildren should be implemented carefully, such as, for example, by undertaking pilot studies of limited scope to evaluate the potential transfer effects of chess instruction to other academic and cognitive fields.
17. Chess instruction has been mainly focused on demonstrating its impact on academic and cognitive areas. It would also be worthwhile, though, to evaluate the impact of chess training on other aspects of human

behaviour, such as the self-regulation of emotions, or adjustment to social and school circumstances.

18. Most of the studies about chess psychology have been possible because chess players have been willing to participate in them. Without a good disposition to participate in the scientific pursuit about the psychology of chess, none of the studies examined in this book would have been feasible. This last point is, therefore, a big 'thank you' to all the chess players who have aided in the investigation of the issues addressed in this book.

This book compiles most, if not all, of the relatively extensive body of behavioural research that uses chess as a model domain. The main emphasis has been placed, however, on describing what the emerging individual differences are that have been addressed in this field of research. These individual differences range from research into the biological basis of behaviour to the study of the most common traits and central themes usually addressed in differential psychology. Chess is a very simple but universal domain with very clear rules. The chess skill of millions of chess players is accurately gauged with the Elo chess rating. These appealing features are indicative of the fact that chess can be adapted to a variety of paradigms and research programmes within the broader field of individual differences.

# APPENDIX 1

Studies with the Elo rating within psychological research ($n$ = 134). Superscripts next to the sample sizes ($N$): cg = control group; eg = experimental group; co = coaching; nc = no coaching; hp = high practice; lp = low practice; ex = expert chess players; no = novice chess players; re = recreational chess players

| Study | Journal | N | Main conclusion(s) |
| --- | --- | --- | --- |
| (Amidzic et al., 2006) | *Journal of Psychophysiology* | 20 | Time-dependent reorganization during the formation of expert memory can be studied in humans and support the theory that the medial temporal lobe and hippocampal formation play a transitional role during the creation of expert memory in the neocortex. |
| (Amidzic et al., 2001) | *Nature* | 20 | Amateur players show higher activity in the medial temporal lobe. Grandmasters show more activity in the frontal and parietal cortices. |
| (Avni et al., 1987) | *Personality and Individual Differences* | $20^{ex}$<br>$20^{no}$<br>20 | Highly competitive players differed from non-players in being significantly more suspicious. |
| (Bachmann & Oit, 1992) | *Psychological Research* | $18^{ex}$<br>$14^{no}$<br>14 | Imagery seems to constitute a dynamic process of interplay between visuospatial and verbal-propositional codes. |
| (Berthelot et al., 2012) | *Age* | $96^{ex}$ | There is an exponential biphasic growth and decline process concerning chess performance. |

| | | | |
|---|---|---|---|
| (Bertoni et al., 2015) | *Journal of Economic Behavior and Organization* | 40,530 | Less talented players are more likely to drop out. The age-productivity gradient is heterogeneous by ability, making fixed effects estimators inconsistent. |
| (Bilalić et al., 2010) | *Journal of Experimental Psychology: General* | $8^{ex}$<br>$15^{no}$ | The superiority of expert players in object recognition relates to bilateral activity next to the occipito-temporal junction. Experts have an extensive and developed knowledge of domain-specific patterns, although their performance drops with the randomization of usual object location. |
| (Bilalić et al., 2007a) | *Intelligence* | 57 | Practice but not intelligence is associated with chess skill. Children who are more intelligent dedicate more time to chess. In a chess elite subsample, less intelligent children dedicate more time to chess. |
| (Bilalić et al., 2008a) | *Cognitive Psychology* | 58<br>36<br>12 | Expertise and thought flexibility are positively associated. There is also inflexibility among ordinary experts in front of super-experts. |
| (Bilalić et al., 2008b) | *Cognition* | $10^{ex}$<br>$46^{ex}$ | More time is invested on the relevant squares for familiar solutions than on the squares relevant for the optimal solution. One fixed idea prevents others coming to mind. |
| (Bilalić, McLeod, et al., 2009) | *Cognitive Science* | 24 | There is a link in experts between problem solving and memory of specific experiences. Context-independent general-purpose problem-solving strategies to teach future experts are unlikely to succeed. |

*Cont.*

| Study | Journal | N | Main conclusion(s) |
|---|---|---|---|
| (Bilalić, Smallbone, et al., 2009) | *Proceedings of the Royal Society* | 200 | The performance of the best German male chess players is better than that of the best German female players. The underrepresentation of women in chess is due to statistical sampling, not to sex differences in intellectual ability. |
| (Bilalić, Kiesel, et al., 2011) | *PLOS ONE* | $8^{ex}$<br>$8^{no}$ | Experts' extensive knowledge of domain objects and their functions enables superior recognition even when experts are not directly fixating the objects of interest. Functional magnetic resonance imaging (fMRI) relates exclusively the areas along the dorsal stream to chess-specific object recognition. |
| (Bilalić, Langner, et al., 2011) | *Journal of Neuroscience* | $7^{ex}$, $8^{no}$<br>$7^{ex}$, $7^{no}$<br>$6^{ex}$, $7^{no}$ | Fusiform face area (FFA) activity relates to stimulus properties and not to chess skill directly. In all chess and non-chess tasks, experts' FFAs are more activated than those of novices only when they deal with naturalistic full-board chess positions. |
| (Bilalić et al., 2012) | *Human Brain Mapping* | $8^{ex}$<br>$15^{no}$ | Experts are faster than novices on the enumeration of relations between chess objects because their extensive knowledge enables them to immediately focus on the objects of interest. Expert-like cognitive processing can generalize laboratory results to everyday life. |

| | | | |
|---|---|---|---|
| (Blanch, 2016) | *Personality and Individual Differences* | 112,358 | There is a male advantage in chess performance throughout twenty-four countries. Sex differences in chess performance emerge for all the studied countries, with remarkable and highly variable unexplained gaps unrelated to the men versus women ratios. |
| (Blanch, 2018) | *Intelligence* | 100 | The dynamic change in tournament activity is a stronger predictor of the change in performance than the static cumulated amount of activity. Age and the change in tournament activity explain only half the variation in the change in performance across the observed period. |
| (Blanch et al., 2015) | *Personality and Individual Differences* | 553 | Sex differences in Elo ratings are unrelated to discrepant participation rates. Age and practice predict sex differences in Elo ratings for women. |
| (Blanch, García, Llaveria, et al., 2017) | *Personality and Individual Differences* | 259 | A structural equation model with five subtests from the ACT for low and high Elo chess rating groups indicates a better overall model fit for the low-ability group, supporting Spearman's law of diminishing returns. |
| (Boggan et al., 2012) | *Journal of Experimental Psychology: General* | $27^{ex}$ $22^{re}$ $20^{no}$ | At a high experience level, face and object recognition share common processes. Chess expertise relates positively to a congruency effect with chess, but negatively to a congruency effect with faces. |

*Cont.*

| Study | Journal | N | Main conclusion(s) |
|---|---|---|---|
| (Breznik & Law, 2016) | *Perceptual and Motor Skills* | 26,547 | There is a relative age effect (RAE) among top chess players. In general, a winter/spring excess of birth dates is observed in both males and females. There are differences in RAE between genders and age groups. There is a reverse RAE in the male category. |
| (Bühren & Frank, 2012) | *Talent Development and Excellence* | 6,112 | Chess grandmasters act very similarly to other humans. This even holds true when they play exclusively against players of approximately their own strength. Their skills are rather specific for their game. |
| (Burgoyne et al., 2016) | *Intelligence* | – | Chess skill correlates positively with fluid reasoning, comprehension/knowledge, short-term memory, and processing speed. Chess skill correlates more strongly with numerical ability than with verbal ability or visuospatial ability. Cognitive ability contributes to individual differences in chess skill, particularly in young chess players and/or at lower levels of skill. |

| | | | |
|---|---|---|---|
| (Burns, 2004) | *Psychological Science* | 348<br>584<br>247 | Performance in blitz chess is strongly associated with performance in non-blitz chess. Chess skill differences equalize in blitz chess, though to a much lesser extent for highly skilled players. Chess search skill does not improve much as players' skill levels rise. |
| (Calderwood et al., 1988) | *American Journal of Psychology* | 6 | Blitz time constraints have a greater detrimental effect on move quality for weaker than for stronger players. Skilled players rely more on rapid recognition processes. |
| (Campitelli & Gobet, 2004) | *Journal of International Computer Games Association* | 4 | The ability to search and pattern recognition are relevant aspects of expert thinking. Long-term memory knowledge allows extensive search and rapid evaluation when making decisions under time pressure. Players adaptively use either problem-solving method depending on the demands of the task. |
| (Campitelli et al., 2005) | *Spanish Journal of Psychology* | 16 | There is brain activation in the frontal areas of the novices but not in the masters, who, rather, use from anterior to posterior areas of the brain. The difference between masters and novices relies on the creation of long-term memory with typical chess configurations. |
| (Campitelli & Gobet, 2005) | *European Journal of Cognitive Psychology* | 16<br>16 | Irrelevant information affects chess performance only when it changes during the presentation of a given game. |

*Cont.*

| Study | Journal | N | Main conclusion(s) |
|---|---|---|---|
| (Campitelli & Gobet, 2008) | *Learning and Individual Differences* | 104 | Practice is more beneficial for improving chess skill at early than at later ages. Group practice is more beneficial than individual practice. Reading books is advisable only if complemented by interaction with the environment. |
| (Campitelli & Gobet, 2011) | *Current Directions in Psychological Science* | – | Deliberate practice is necessary but not sufficient to achieve a high level of chess performance. Cognitive abilities, early starting age, handedness, and season of birth also influence chess performance. |
| (Campitelli et al., 2007) | *International Journal of Neuroscience* | 5 | Chess players access long-term memory chunks of domain-specific information, presumably stored in the temporal lobes. The recognition memory tasks activate working memory areas in the frontal and parietal lobes. |
| (Campitelli et al., 2008) | *International Journal of Neuroscience* | 2[ex] | There is a left-lateralized pattern of brain activity very similar in both masters. The brain areas activated are the left temporo-parietal junction and left frontal areas. |
| (Chabris & Hamilton, 1992) | *Neuropsychologia* | 16 | The right hemisphere is better able to acquire and apply new sets of default parsing rules for specific contexts. The right hemisphere is critical for chess skill. |

| | | | |
|---|---|---|---|
| (Chabris & Glickman, 2006) | *Psychological Science* | 256,741 | There are fewer women than men at the highest level in chess because fewer women enter competitive chess at the lowest level. Understanding why more boys than girls enter competitive chess is a challenging undertaking. |
| (Chabris & Hearst, 2003) | *Cognitive Science* | 23 | Grandmasters play equally well whether or not they can actually see the changing board positions. |
| (Chang & Lane, 2016) | *American Journal of Psychology* | 200 | Stronger players are more likely than weaker players to spend considerable time on a few moves. The stronger players search more deeply and more accurately. Speed chess also allows players to calculate extensively, calling into question the view that the high correlation between speed chess and standard chess supports recognition action theory. |
| (Charness, 1976) | *Journal of Experimental Psychology: Human Learning and Memory* | 12<br>12<br>12<br>12 | Class A and class C players store chess information within the five available seconds irrespective of the interfering tasks. Class A players retain between 10 and 15% more information than class C players do. |
| (Charness, 1981) | *Journal of Experimental Psychology: General* | 34 | Older players match the problem-solving performance of equivalently skilled younger players despite constraints in encoding and information retrieval. Older players take less time to select optimal moves because of more refined move generation processes. |

*Cont.*

| Study | Journal | N | Main conclusion(s) |
|---|---|---|---|
| (Charness et al., 2001) | *Memory and Cognition* | 24 | Stronger players are faster and more accurate than intermediate players are in choosing the best chess move, and make fewer eye fixations per trial and greater-amplitude saccades, albeit with similar eye fixation duration, than intermediate players do. Expert players encode chess configurations rather than individual pieces. |
| (Charness et al., 2005) | *Applied Cognitive Psychology* | 239<br>180 | Practice is critical for chess expertise. Studying chess is as important as playing to achieve title-level success. |
| (Chase & Simon, 1973) | *Cognitive Psychology* | 3 | Stronger players show a higher ability when encoding a variety of chess positions into larger perceptual chunks. |
| (Connors et al., 2011) | *Cognitive Science* | 22 | Chess masters have faster and deeper searching abilities than intermediate players do. Pattern recognition is critical to chess skill and expert performance. |
| (Cooke et al., 1993) | *American Journal of Psychology* | 6<br>8<br>6 | High-level verbal information supports better recall performance of the position. High-level chess knowledge plays a role in skilled chess memory. |
| (De Bruin et al., 2007b) | *Journal of Sport and Exercise Psychology* | 81 | Motivation to engage in deliberate practice relates to the will to achieve exceptional levels of performance. Higher levels of competitiveness and will to excel predict engagement in deliberate practice activities. |

| | | | |
|---|---|---|---|
| (De Bruin et al., 2008) | *British Journal of Psychology* | 81 | Deliberate practice contributes strongly to chess performance, while gender has a weaker effect. |
| (Doll & Mayr, 1987) | *Psychologische Beiträge* | 33 | Chess players score higher than a reference group in the scales of information-processing capacity for complex information, numerical thinking, working speed, and in the CFT3. Intelligence scores are uncorrelated with success in playing chess; one-year changes in the Elo scores do correlate significantly. |
| (Draper, 1963) | *Journal of the Royal Statistical Society* | 23 | The mental capabilities of professional chess players are not adversely affected by advancing age. As players grow older, their playing peculiarities become known, and thus they find it more difficult to defeat others of equal or lesser strength and may sustain more losses than before to better and/or younger players. |
| (Dreber et al., 2013) | *Journal of Economic Behavior and Organization* | 48,234<br>2,000 | Male players choose significantly riskier strategies when playing against an attractive female opponent, though their performance does not improve. Women's strategies are not affected by the attractiveness of the opponent. |

*Cont.*

| Study | Journal | N | Main conclusion(s) |
|---|---|---|---|
| (Duan et al., 2012) | *NeuroImage* | $15^{ex}$ $15^{no}$ | There are anatomical changes in the bilateral caudate of expert players in response to long-term chess training. Involvement in chess problem solving relates to smaller caudate nuclei, enhancing better integration of cognitive skills acquisition to address the resolution of complex problems. |
| (Fair, 2007) | *Experimental Aging Research* | 10 | Chess shows a much slower decline in performance than physical activities such as running and swimming. |
| (Fernandez-Slezak & Sigman, 2012) | *Journal of Experimental Psychology: General* | – | Players adopt a conservative strategy (prevention mode) when facing strong opponents, which translates into slower and more accurate moves. Players are more likely to win against strong opponents with the mode adopted when playing against opponents with similar chess strength. |

| | | | |
|---|---|---|---|
| (Ferrari et al., 2008) | *Quarterly Journal of Experimental Psychology* | $20^{ex}$<br>$20^{no}$<br>$40^{ex}$<br>$40^{no}$ | Experienced players detect changes in undeveloped patterns later. For novices, the detection of changes occurs later than for experienced players (between four and ten seconds), but there is no difference between undeveloped and developed patterns. Experienced players do an initial fast exploration of undeveloped patterns at the periphery of the chessboard, followed rapidly by the encoding of central developed patterns. During the very first seconds of chess-position exploration, experienced players encode undeveloped patterns. |
| (Ferreira & Palhares, 2008) | *The Mathematics Enthusiast* | 437 | There is a relation between strength of play and patterns involving problem solving. |
| (Franke et al., 2017) | *European Neuropsychopharmacology* | 39 | Modifying effects of stimulants on complex cognitive tasks result from more reflective decision-making processes. Without time pressure, such effects may result in enhanced performance. Under time constraints more reflective decision-making may not improve or even have detrimental effects on complex task performance. |
| (Frey & Adesman, 1976) | *Memory and Cognition* | 13<br>27<br>20 | Memory for briefly exposed chess positions is a function of chess skill. Expert players chunk complex visual inputs into familiar groupings, and discover more semantic relations between the different pieces at play and chunks. |

*Cont.*

| Study | Journal | N | Main conclusion(s) |
|---|---|---|---|
| (Frydman & Lynn, 1992) | *British Journal of Psychology* | 33 | High levels of general intelligence and spatial ability are needed to achieve a high chess level. The high spatial ability of young chess players may explain why males tend to be more numerous than females among high-standard players. |
| (Gaschler et al., 2014) | *Frontiers in Psychology* | 1,383 | Negative exponential learning curves explain better the early performance gains and predict later trajectories in chess expertise than power function learning curves. |
| (Gerdes & Gränsmark, 2010) | *Labour Economics* | 15,122 | Women choose more cautious strategies than men do. Men and women are more prone to risky strategies against female opponents who are stronger players. |
| (Gobet, 1992) | *Psychological Research* | 48 | Learned helplessness can be induced by manipulating cognitive contingencies. The development of learned helplessness requires a high similarity between the no contingency learning situation and the following situations, and an average skill in the particular domain. |

| | | | |
|---|---|---|---|
| (Gobet & Campitelli, 2007) | *Developmental Psychology* | 104 | Left-handedness is significantly present in the chess sample. Handedness and skill level are not related. Key factors for achieving a high level of chess expertise are starting to play seriously below the age of twelve, individual and group practice, and feedback from a coach. |
| (Gobet & Chassy, 2008) | *Journal of Biosocial Science* | 42,650 | The population of expert chess players in the Northern Hemisphere shows a seasonal pattern, with an excess of births in late winter and early spring. |
| (Gobet & Clarkson, 2004) | Memory | 12 | The original chunking theory overestimates short-term memory (STM) capacity and underestimates the size of chunks used, in particular with masters. The proposal that STM holds four chunks may be an overestimate. |
| (Gobet & Ereku, 2014) | *Frontiers in Psychology* | 10[ex] | The case of Magnus Carlsen demonstrates that deliberate practice is necessary, but not sufficient, for achieving high levels of expert performance. |
| (Gobet & Simon, 1996b) | *Psychological Science* | 1 | Recognition of chess patterns, based on deeper chess knowledge, plays a much stronger role in simultaneous time-constrained chess than in planning processes. |
| (Gobet & Simon, 1996c) | *Cognitive Psychology* | 13<br>5<br>1 | With time enough, expert players recall chess positions from distinct templates already stored in long-term memory. |

*Cont.*

| Study | Journal | N | Main conclusion(s) |
|---|---|---|---|
| (Gobet & Simon, 1996a) | *Memory and Cognition* | 25 | Chess players' memory diminishes when exposed to mirror-image reflections of chess positions. Expert players' memory chunks are larger than past estimations. Chess players can find chunks within random positions. The outcomes support the template theory. |
| (Gobet & Simon, 1998a) | *Memory* | 26 | Chess expertise depends on the availability in memory of information about recurrent patterns of pieces, and highly selective search strategies within the move tree. |
| (Gobet & Simon, 2000) | *Cognitive Science* | 21 | Stronger players are better at recalling both 'normal' and random chess positions. Templates help to explain their better performance even with shorter presentation times. |
| (Gobet & Waters, 2003) | *Journal of Experimental Psychology: Learning, Memory, and Cognition* | 36 | There is a chess skill effect with random location and distribution of the pieces on a chessboard. The template theory of expert memory provides a plausible account for expert memory in chess. |
| (Goldin, 1978) | *American Journal of Psychology* | 8<br>9 | An orienting task that requires the encoding of meaningful aspects of a position results in better recognition than from a more structural task. Aspects of meaning have some input into processes that generate the memory representation of chess positions. |

| | | | |
|---|---|---|---|
| (Goldin, 1979) | *American Journal of Psychology* | 7<br>8 | Chess strength influences strongly recognition accuracy. Analyses of chess positions form a component of the memory representation, particularly for stronger players. Recognition and recall tasks tap the same basic representational structures. |
| (Gong et al., 2015) | *Plos One* | 25<br>25[re] | For recall of game positions, significant differences between players and non-players are found in virtually all the characteristics of chunks recalled. The size of the largest chunks also correlates with chess skill within the group of rated chess players. |
| (Grabner et al., 2006) | *Brain Research Bulletin* | 47 | Intelligence and chess expertise have a different impact on neural efficiency. More skilled chess players display higher activation over the parietal cortices and lower activation over the frontal cortices in the speed and reasoning tasks. |
| (Grabner et al., 2007) | *Acta Psychologica* | 90 | Playing strength relates more strongly to numerical aptitudes than to verbal or visuospatial aptitudes. Chess skill relates strongly to chess experience. The big five personality dimensions do not relate to chess skill; emotion expression control does. |

*Cont.*

| Study | Journal | N | Main conclusion(s) |
| --- | --- | --- | --- |
| (Gränsmark, 2012) | *Journal of Economic Behavior and Organization* | 30,000<br>1,620 | Men perform worse than women in shorter games but better in longer games. Women perform worse under time pressure. Men are more inconsistent because they tend to be too impatient. Women are more inconsistent as they tend to over-consume reflection time in the beginning, leading to time pressure later. |
| (Hanggi et al., 2014) | *Neuropsychologia* | $20^{ex}$<br>$20^{cg}$ | There are differences in grey and white matter morphology between chess players and control subjects in brain regions associated with cognitive functions important for playing chess. Whether these anatomical alterations are the cause or consequence of the intensive and long-term chess training and practice is unknown. |
| (Holding & Pfau, 1985) | *American Journal of Psychology* | 24 | Stronger players produce moves with a higher quality, due to their higher evaluative judgement abilities. Chess players at different levels of skill show differences in cognitive processing apart from memory. Stronger players have smaller forward-visual evaluation discrepancies. |
| (Holding & Reynolds, 1982) | *Memory and Cognition* | 24 | The quality of the chosen moves is independent of recall performance. The selected moves are not based on the few remembered pieces. |

| | | | |
|---|---|---|---|
| (Horgan et al., 1989) | *International Journal of Personal Construct Psychology* | 6<br>35 | Expert players show more sophisticated and general similarity judgement strategies than novice players do. |
| | | 35<br>12 | Expert players agree more often about candidate moves, and generate them more quickly than novice players do. There are relevant individual differences in information-processing strategies applied by expert players. |
| (Horgan & Morgan, 1990) | *Applied Cognitive Psychology* | 113<br>15 | Chess club activities contribute to improve chess skill. Spatial and logical abilities serve to identify chess talent. |
| (Horgan, 1992) | *Psychological Research* | 83 | Better players make better predictions, while worse players tend to be overconfident. Players with better predictions know what they need to work. |
| | | 59<br>35<br>47 | Skill level is associated with the number of pieces correctly remembered. Overall, good training is sufficient to produce good players above chess-specific talent. |
| (Howard, 1999) | *Intelligence* | 50 | Younger players have dominated the game since the 1970s, highlighting the fact that there has been a general rise in the intelligence of the population. |

*Cont.*

| Study | Journal | N | Main conclusion(s) |
| --- | --- | --- | --- |
| (Howard, 2005b) | Personality and Individual Differences | 660<br>160<br>1,351<br>60 | Since 1970 chess players have achieved high chess performance levels at younger ages. This effect occurs more within top players, and, to a lesser extent, within top female players. The age effect is unlikely to be due to chess environment changes but to natural talent, consistent with the rising population ability or Flynn effect. |
| (Howard, 2009) | *Memory and Cognition* | 3,471<br>69,998<br>265<br>327<br>5 | The expertise development pattern follows a logarithmic function that reaches asymptote after extensive actual practice. Top players show signs of natural talent – precocity, faster acquisition of expertise, and a higher peak performance level – after extensive actual practice. |
| (Howard, 2011b) | *Behavior Research Methods* | 668 | Recall of entry year in chess and games played in one year are accurate on average. There are players who are very inaccurate in recalling entry year and total games. |
| (Howard, 2011a) | *Cognitive Development* | 12 | Practice differences are unable to explain chess performance differences between the Polgar sisters. Innate talent is a contributor to chess performance. |
| (Howard, 2012a) | *Applied Cognitive Psychology* | 533 | The number of games played is the strongest predictor of chess rating. The number of study hours is also a predictor of chess rating, but formal coaching is not. |

| | | | |
|---|---|---|---|
| | | $385^{co}$<br>$288^{nc}$ | Coaching has a longitudinal effect on chess rating, even though it may take time to become apparent. |
| | | $39^{hp}$<br>$38^{lp}$ | The number of study hours is a weak predictor of chess rating. Its impact is low when controlling for the number of games and time remaining in the domain of chess. |
| | | 202<br>94<br>97<br>31<br>24 | The number of games played is a powerful predictor of chess rating, holding domain persistence constant. Natural talent affects chess expertise development, however. |
| (Howard, 2013) | *British Journal of Psychology* | 347<br>74<br>680 | The number of rated games is the strongest predictor of peak rating. The number of chess total study hours is a weaker predictor. Chess total study hours are not a predictor of chess rating at 350 games. Reaching a performance ceiling and undertaking specific practice to go beyond it happens independently of player strength. Many players play few internationally rated games because of involvement in work and education activities. Extensive study is not a major causal factor in chess performance. |

*Cont.*

| Study | Journal | N | Main conclusion(s) |
|---|---|---|---|
| (Howard, 2014c) | *Acta Psychologica* | 387<br>38 | A power or logarithmic function best fits group data but individuals show much variability. An exponential function usually is the worst fit to individual data. A power function best fits the group curve for the more talented players while a quadratic function best fits that for the less talented. Individual variability is great and the power or exponential laws are not the best descriptions of individual chess skill development. |
| (Howard, 2014a) | *Personality and Individual Differences* | 139,000 | The sex difference in performance in the top ten and fifty of all international players is large at about one standard deviation and stayed roughly constant from 1975 to 2014. Male predominance in chess and related domains may be due partly to sex differences in innate abilities. |
| (Howard, 2014b) | *Journal of Biosocial Science* | 1,203<br>21<br>230<br>85,832<br>535 | Males are very disproportionately represented in the top chess rankings. Top female players show signs of having less natural talent for chess than top males, such as taking more rated games to gain the grandmaster title. Males on average may have some innate advantages in developing and exercising chess skill. |

| | | | |
|---|---|---|---|
| (Jastrzembski et al., 2006) | *Psychology and Aging* | $20^{ex}$<br>$19^{no}$<br>$20^{re}$ | Expert chess players have a perceptual advantage over less skilled players, a critical component of chess expertise that differentiates level of skill. Basic abilities are not good predictors of skill in chess. Knowledge activation processes to assess basic chess relationships slow with age even in expert chess players. |
| (Kelly, 1985) | *Journal of Personality Assessment* | 270<br>$209^{ex}$ | Chess players possess different personality and temperamental characteristics. Chess masters also vary in personality and temperament compared with average players. Intuition is the personality factor most related to chess skill, and the dominant process that distinguishes master from average players. Chess may be more a game based on intuition than on logical thinking. |
| (Kiesel et al., 2009) | *Journal of Experimental Psychology: Learning, Memory, and Cognition* | $12^{ex}$<br>$24^{no}$<br>$12^{ex}$ | Expert chess players show higher perceptual skills than novice players do. Chess experts process chess configurations even when they are presented unconsciously. A high amount of practice is necessary to attain this level of unconscious information processing. |
| (Klein & Peio, 1989) | *American Journal of Psychology* | 34 | Stronger players are more accurate than novices in predicting the moves of other proficient players. Stronger players are more likely to generate the correct option as the first one considered. |

*Cont.*

| Study | Journal | N | Main conclusion(s) |
|---|---|---|---|
| (Klein et al., 1995) | *Organizational Behavior and Human Decision Processes* | $16^{ex}$ | Skilled decision-makers pursue deep rather than broad searches, relying on their ability to generate satisfactory options as the first ones considered. |
| (Knapp, 2010) | *Proceedings of the Royal Society* | 200 | The conclusion of Bilalić, Smallbone, et al. (2009) that 'there is little left for biological or cultural explanations to account for' appears to be premature. |
| (Krawczyk et al., 2011) | *Neuroscience Letters* | $6^{ex}$ $6^{no}$ | The configurations in chess are not strongly processed by face-selective regions that are selective for faces in individuals who have expertise in both domains. The area most involved in chess does not show overlap with faces. Expert visual processing may be similar at the level of recognition, but need not show the same neural correlates. |
| (Lane & Robertson, 1979) | *Memory and Cognition* | $19^{no}$ $16^{ex}$ | Novice and expert players generate long-lasting representations of chess positions very quickly. Expert players are better than novice players are at integrating familiar configurations into a coherent whole, and are familiar with more patterns of chess pieces. |

| | | | |
|---|---|---|---|
| (Leone et al., 2012) | *Frontiers in Human Neuroscience* | 9 | Heart rate signal events predict the conception of a plan, the concrete analysis of variations, or the likelihood to blunder by fluctuations before the move, and they reflect reactions, such as a blunder made by the opponent, by fluctuations subsequent to the move. |
| (Levitt et al., 2011) | *American Economic Review* | 206 | The centipede results resemble the empirical findings with subjects from the general population, whereby stopping at the first node is rare despite being the best decision (Nash equilibrium). Chess skill is unrelated to performance in the game race to 100. |
| (Mazur et al., 1992) | *Social Psychology Quarterly* | 16[eg] | Winners of chess tournaments show higher testosterone levels than losers do. Sometimes competitors show rises in testosterone before their games, as if in preparation. |
| (McGregor & Howes, 2002) | *Memory and Cognition* | 24<br>24 | Stronger players determine long-term memory chunks of chess information from attack/defence relations between the intervening pieces. |
| (Moxley et al., 2012) | *Cognition* | 37[ex]<br>34[re] | Expert and less skilled chess players benefit from deliberative thinking when facing different types of chess problems. Experts rely consistently on deliberate search to improve the quality of move selection. |

*Cont.*

| Study | Journal | N | Main conclusion(s) |
|---|---|---|---|
| (Moxley & Charness, 2013) | *Psychonomic Bulletin Review* | – | Ageing affects performance on the best move choice task. The recall task is an effective measure of chess skill, and it is effective to study age and skill differences. |
| (Onofrj et al., 1995) | *Neuroscience Letters* | 5 | The non-dominant frontal lobe is active in the brain of chess experts when elaborating a solution for a complex chess problem. Chunking of elements into meaningful groupings and parsing of visual stimuli are functions of the non-dominant hemisphere. |
| (Palacios-Huerta & Volij, 2009) | *American Economic Review* | 422<br>40[cg] | Chess players tend to play very differently from college students. A significant majority of chess players choose the only action that is consistent with equilibrium. |
| (Pfau & Murphy, 1988) | *American Journal of Psychology* | 59 | Pattern recognition memory interacts with chess verbal knowledge. Chess verbal knowledge is a determinant of chess skill. A wealthy chess verbal network enables a chess master to have a relevant memory level for a variety of chess positions. |
| (Reingold et al., 2001) | *Psychological Science* | 32 | Expert chess players display a higher perceptual encoding ability, attributable to their higher chess experience rather than to a general perceptual or memory superiority. |

| | | | |
|---|---|---|---|
| (Reynolds, 1982) | *American Journal of Psychology* | 9 | Master-level chess players show superior recall for briefly presented positions only when the material affects a highly centralized area of the board. |
| (Reynolds, 1992) | *American Journal of Psychology* | 15 | Stronger players make better estimations of players' strength on the basis of observed chess positions. Estimation error decreases after seeing the subsequent moves in the position. Seeing moves in addition to the initial position is essential for the accurate recognition of playing level. |
| (Robbins et al., 1996) | *Memory and Cognition* | 20<br>12<br>15 | Verbal rehearsal plays a weak role in memory for chess positions or best chess move selection. Stronger players process chess information more efficiently than weaker players do. The selection of moves depends on the capacity for representations to elicit candidate moves. |
| (Roring & Charness, 2007) | *Psychology and Aging* | 5,011 | The age of peak performance occurs later in life than originally proposed and is independent of the initial skill level. Ageing is slightly kinder to the initially more able, who show milder declines past their peak. Higher tournament activity levels predict higher ratings and interact with age in the initially more able sample. Activity has smaller effects on rating for older adults. |

*Cont.*

| Study | Journal | N | Main conclusion(s) |
|---|---|---|---|
| (Rothgerber & Wolsiefer, 2014) | *Group Processes and Intergroup Relations* | 414 | Females perform worse than expected when playing against a male opponent. Stereotype threat is most pronounced when playing a strong or moderate opponent and when playing someone in a higher or the same grade. Those most vulnerable to stereotype threat are less likely to continue playing in future chess tournaments. |
| (Ruiz & Luciano, 2009) | *Psicothema* | 20 | Chess performance improves after an intervention based on acceptance and commitment therapy (ACT). |
| (Saariluoma, 1985) | *Memory and Cognition* | 14<br>6<br>9 | Stronger players are faster in recalling chess positions. Random positions are more difficult to recall than game positions. Differences in reaction times across different kinds of positions are independent of chess skill. |
| (Saariluoma & Kalakoski, 1997) | *American Journal of Psychology* | 12<br>6<br>8<br>6 | Blindfold chess imagery formation is independent of the modality of presented information. Stronger players are superior in encoding speed and accuracy than medium-level players. Skilled imagery builds on long-term working memory retrieval structures. The construction of complex mental images needs visual working memory. |

| (Saariluoma et al., 2004) | *European Journal of Cognitive Psychology* | 6 | In expert players, chess images are different from ordinary mental images. Visuospatial representations are characterized by large learned chunks and automated processing habits. |
|---|---|---|---|
| (Sala, Burgoyne, et al., 2017) | *Intelligence* | – | Chess players outperform non-chess players in intelligence-related skills. Access to training alone cannot explain expert performance. |
| (Schultetus & Charness, 1999) | *American Journal of Psychology* | 17 | There is a clear relationship between skill, the quality of moves chosen, and the ability to recall both game and quasi-random positions. Pattern recognition processes underlie superior performance by skilled chess players. |
| (Schwarz et al., 2003) | *Psychosomatic Medicine* | 9[ex] | Increments in helplessness/hopelessness are associated with reduced high-frequency heart rate variability (HF-HRV). Negative mood states and autonomic nervous system disturbances are related to cardiac events. |
| (Sigman et al., 2010) | *Frontiers in Neuroscience* | 44,069 (2,067 computers) | The capacity of blunders and score fluctuations are characterized to predict player strength, which is still an open problem in chess software. The winning likelihood can be reliably estimated from a weighted combination of remaining times and position evaluation. |

*Cont.*

| Study | Journal | N | Main conclusion(s) |
|---|---|---|---|
| (Stafford, 2018) | *Psychological Science* | 461,637 | Female players outperform expectations when playing men. There is no influence from degree of challenge, player age, nor prevalence of female role models in national chess leagues on differences in performance when women play men versus when they play women. This analysis contradicts the influence of gender stereotypes. The differences between male and female players suggest that systematic factors do exist and remain to be uncovered. |
| (Unterrainer et al., 2006) | *British Journal of Psychology* | 50 | Chess players outperform non-chess players in planning performance, increasing this difference with more difficult problems. There are no differences in fluid intelligence or working memory between both groups. |
| (Unterrainer et al., 2011) | *American Journal of Psychology* | 30 | Chess players report higher overall trait and state motivation scores across both experiments. There is no general transfer of chess-related planning expertise to other cognitive domains. Superior performance may be possible only under specific circumstances, such as receiving competitive instructions. |

| | | | |
|---|---|---|---|
| (Vaci et al., 2015) | *Psychology and Aging* | 100,529 | After reaching the peak at around thirty-eight years, the more able players deteriorate more quickly. Their decline starts to slow down at around fifty-two years, earlier than for less able players (fifty-seven years). The decline and its stabilization are significantly influenced by activity. The more players engage in playing tournaments, the less they decline and the earlier they start to stabilize. |
| (Van der Maas & Wagenmakers, 2005) | *American Journal of Psychology* | 259 | The Amsterdam Chess Test is a valid and reliable test to assess chess expertise in a limited time. |
| (Van Harreveld et al., 2007) | *Psychological Research* | 300 | Correlations of ICC ratings with different time controls are higher for weaker players, and lower for stronger players. Chess skill is less predictive of performance in blitz than in slower games. |
| (Vasyukova, 2012) | *Psychology in Russia: State of the Art* | 44 | Verbalized operational senses (VOSs) depend on the factors of chess position and the age and skill level of the player. VOS transfer is found more in skilled than in unskilled chess players. Skilled players demonstrate selectivity of search in a connected position. VOS transfer is associated with using and transforming the results of previous verbal searches. |

*Cont.*

| Study | Journal | N | Main conclusion(s) |
|---|---|---|---|
| (Volke et al., 2002) | *Journal of Psychophysiology* | $11^{ex}$<br>$11^{no}$ | Experts show well-mastered and automatic performance in terms of the relationships of different cortical areas. Novices show poorly mastered tasks. |
| (Vollstädt-Klein et al., 2010) | *Learning and Individual Differences* | 40 | The importance of personality factors for chess skill differs by gender. Male players are more introverted than female players. |
| (Waters & Gobet, 2008) | *Memory and Cognition* | 36<br>20 | The CHREST computational model is accurate for explaining mental imagery in chess. |
| (Waters et al., 2002) | *British Journal of Psychology* | 36 | Visual memory ability is not associated with chess skill. Visuospatial abilities might be relatively unimportant for the acquisition of chess skill in the long term. |
| (Wright et al., 2013) | Psychophysiology | $14^{ex}$<br>$14^{no}$ | Experts show an enhanced N2 and larger P3 in chess tasks. Expert–novice differences in posterior N2 begin early on check-related searches (240 milliseconds). Prolonged N2 components reflect the matching of current perceptual input to memory, being sensitive to experts' superior pattern recognition and memory retrieval of chunks. |

# APPENDIX 2

Cross-country variability in chess skill as measured by the Elo rating (FIDE list, December 2018): the map below represents the mean Elo rating (M = 2188, Sd = 335, coefficient of variation (Sd / M) = 0.15); the map overleaf represents the number of chess grandmasters (M = 9.38, Sd = 24.56, coefficient of variation (Sd / M) = 2.62); countries in white are missing

**Mean Elo ratings**

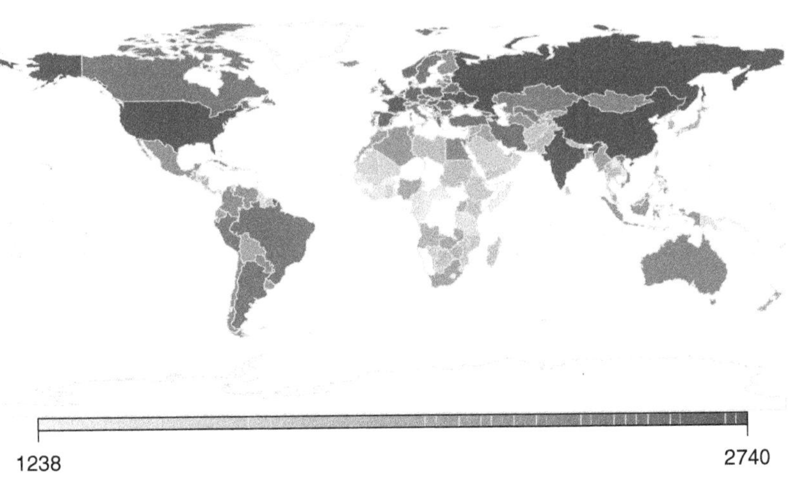

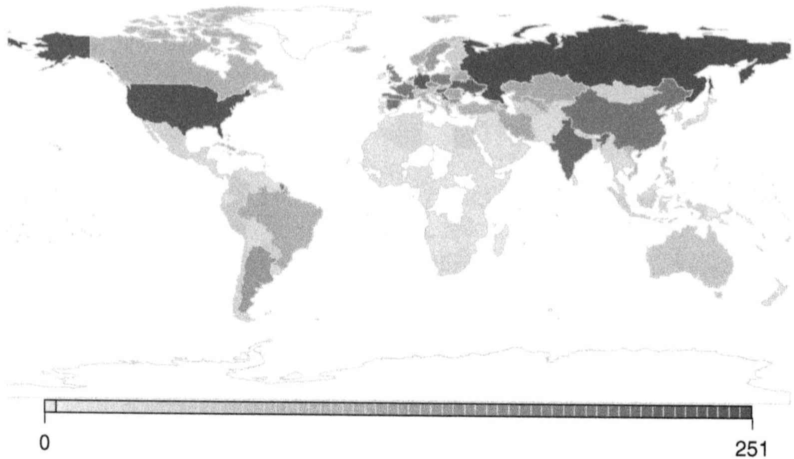
Number of chess grandmasters
0
251

# APPENDIX 3

Studies about chess and intelligence in child and adult samples (BIS = Berlin Structural Model of Intelligence; CFT-3 = Culture Fair Intelligence Test; DAT = Differential Aptitude Test; GATB = General Aptitude Test Battery; IST = Intelligenz-Struktur-Test; M:F = male to female ratio; PMA = Primary Mental Aptitudes; RPM = Raven's Progressive Matrices; ToL = Tower of London; TONI-3 = Test of Nonverbal Intelligence; WCST = Wisconsin Card Sorting Test)

| Study | Sample (ages) | N | Ability | Chess skill | Design | M:F | Country |
|---|---|---|---|---|---|---|---|
| (Aciego et al., 2012) | Children (6–16) | 210 | WISC-R | Problem solving test | Experimental | – | Spain |
| (Aciego et al., 2016) | Children (6–16) | 230 | WISC-R | Chess test | Experimental | – | Spain |
| (Bilalić et al., 2007a) | Children (–) | 57 | WISC III (IQ) | Chess test<br>Elo rating | Correlational | 4:1 | United Kingdom |
| (Blanch, García, Llaveria, et al., 2017) | Mixed (8–70) | 259 | *g* factor | Elo rating | Correlational | 14:1 | Netherlands |
| (De Bruin et al., 2014) | Children (6–11) | 24 | WISC III (IQ) | Chess test | Correlational | 2:1 | Netherlands |
| (Doll & Mayr, 1987) | Adults (18–51) | 33 | BIS, CFT-3 | Elo rating | Correlational | 33:0 | Germany |
| (Duan et al., 2012) | Adults (–) | 30 | RPM | Elo rating | Experimental | 2:1 | China |
| (Frank & D'Hondt, 1979) | Children (–) | 90 | GATB, PMA, DAT | Chess test | Experimental | – | Zaire |
| (Frydman & Lynn, 1992) | Children (8–13) | 33 | WISC | Elo rating | Experimental | 7:1 | Belgium |
| (Gliga & Flesner, 2014) | Children (–) | 38 | Dearborn test | Elo rating | Experimental | 1:1 | Romania |
| (Grabner et al., 2006) | Adults (18–65) | 47 | IST (IQ) | Elo rating | Experimental | 47:0 | Austria |
| (Grabner et al., 2007) | Adults (15–65) | 90 | IST (IQ) | Elo rating | Correlational | 29:1 | Austria |

| | | | | | | | |
|---|---|---|---|---|---|---|---|
| (Grau-Pérez & Moreira, 2017) | Children (7–12) | 28 | RPM, ToL, WCST | – | Experimental | 6:1 | Uruguay |
| (Hanggi et al., 2014) | Adults (–) | 40 | RPM; visuo-spatial abilities | Elo rating | Experimental | 20:0 | Switzerland |
| (Hernández & Rodríguez, 2006) | Children (10–16) | 53 | RPM | Elo rating | Experimental | – | Spain |
| (Hong & Bart, 2007) | Children (–) | 38 | RPM TONI-3 | Elo rating; chess test | Experimental | 2:1 | South Korea |
| (Horgan & Morgan, 1990) | Children (–) | 148 | RPM | Elo rating | Experimental | 13:2 | United States |
| (Jastrzembski et al., 2006) | Adults (17–81) | 59 | Speed of information processing; working memory | Elo rating | Experimental | – | United States |
| (Kazemi et al., 2012) | Children (–) | 180 | Metacognitive | – | Experimental | 180:0 | Iran |

*Cont.*

| Study | Sample (ages) | N | Ability | Chess skill | Design | M:F | Country |
|---|---|---|---|---|---|---|---|
| (Li et al., 2015) | Adults (17–59) | 29 | RPM | Elo rating | Experimental | 2:1 | China |
| (Nippold, 2009) | Children (7–15) | 32 | Language productivity | Chess test | Experimental | 10:1 | United States |
| (Ramos et al., 2018) | Children (8–12) | 65 | Stroop test; WISC IV (working memory); WCST; mazes | – | Experimental | 1:1 | Argentina |
| (Rojas Vidaurreta, 2011) | Children (7–11) | 44 | WCST | Elo rating | Experimental | 44:0 | Cuba |
| (Schneider et al., 1993) | Children (10–13); adults (22–42) | 40<br>40 | Digit span memory test | Elo rating | Experimental | 40:0<br>1:1 | Germany |
| (Scholz et al., 2008) | Children (10–11) | 53 | Calculation | – | Experimental | 2:1 | Germany |
| (Thompson, 2003) | Children (7–11) | 508 | IQ | Elo rating | Correlational | 508:0 | Australia |

| | | | | | | | |
|---|---|---|---|---|---|---|---|
| (Unterrainer et al., 2006) | Adults (–) | 50 | RPM; ToL; working memory | Elo rating | Experimental | 12:1 | Germany |
| (Unterrainer et al., 2011) | Adults (20–50) | 30 | WAIS; ToL | Elo rating | Experimental | 9:1 | Germany |
| (Van der Maas & Wagenmakers, 2005) | Mixed (11–78) | 259 | Visual memory | Elo rating | Correlational | 14:1 | Netherlands |
| (Waters et al., 2002) | Adults (18–47) | 36 | Visual memory | Elo rating | Experimental | 10:1 | United Kingdom |

# APPENDIX 4

Studies about chess and personality (BFQ-C = Big Five Questionnaire for Children; EPQ = Eysenck Personality Questionnaire; FPI-R = Freiburg Personality Inventory – Revised; M:F = male to female ratio; MBTI = Myers–Briggs Type Indicator; md = Median; MMPI = Minnesota Multiphasic Personality Inventory; SSS = Sensation Seeking Scale)

| Study | Sample (Age range) | N | Personality measure and factors | Study Design | M:F | Country |
|---|---|---|---|---|---|---|
| (Aciego et al., 2012) | Children (6–16) | 210 | TAMAI [multifactor self-assessment test of child adjustment] (personal maladjustment, school maladjustment, social maladjustment) | Experimental | – | Spain |
| (Aciego et al., 2016) | Children (6–16) | 230 | TAMAI [multifactor self-assessment test of child adjustment] (personal maladjustment, school maladjustment, social maladjustment) | Experimental | – | Spain |
| (Avni et al., 1987) | Adults (17–65) | 60 | MMPI (infrequency, paranoia, neuroticism, hostility, aggression, compulsiveness) | Experimental | 60:0 | Israel |
| (Bilalić et al., 2007b) | Children (8–11) | 269 | BFQ-C (extraversion, agreeableness, conscientiousness, instability, openness) | Mixed | 2:1 | United Kingdom |
| (Grabner et al., 2006) | Adults (18–65) | 47 | NEO-FFI (neuroticism, extraversion, openness, agreeableness, conscientiousness) | Experimental | 47:0 | Austria |
| (Grabner et al., 2007) | Mixed (15–65) | 90 | NEO-FFI (neuroticism, extraversion, openness, agreeableness, conscientiousness) | Correlational | 29:1 | Austria |

*Cont.*

| Study | Sample (Age range) | N | Personality measure and factors | Study Design | M:F | Country |
|---|---|---|---|---|---|---|
| (Hernández & Rodríguez, 2006) | Children (10–16) | 53 | MOLDES (active-vital involvement framing, adjustment focal framing, optimizing focal framing) | Experimental | – | Spain |
| (Joireman et al., 2002) | Adults (md = 19) | 82 | SSS (disinhibition, experience seeking, thrill and adventure seeking) | Correlational | – | United States |
| (Kelly, 1985) | Adults (–) | 479 | MBTI (temperament [extraversion, introversion, sensing, intuition, thinking, feeling, judging, perceiving], personality characteristics [sensing-thinking, sensing-feeling, intuitive-thinking, intuitive-feeling], dominant processes [sensing, intuition, thinking, feeling] | Mixed | 3:1 | United States |
| (Llaveria et al., 2016) | Adults (18–77) | 100 | EPQ (extraversion, neuroticism, psychoticism) | Mixed | 37:1 | Spain |
| (Vollstädt-Klein et al., 2010) | Adults (18–63) | 40 | FPI-R (life satisfaction, social orientation, achievement orientation, inhibition, excitability, aggressiveness, stress, somatic complaints, health concerns, frankness, extraversion, emotionality) | Mixed | 3:1 | Germany |

# GLOSSARY

**ATI (aptitude by treatment interaction):** an instructional theory that underlines the importance of the interaction of individual differences in cognitive abilities with different kinds of instructional strategies in producing learning outcomes.

**Backward induction:** the reasoning process that begins at the end of a given problem and attempts to determine the logical sequence that led to that situation.

**Big five personality model:** modelization of the structure of human personality that has gained a considerable consensus. This structure comprises five broad factors: neuroticism, extraversion, openness to experience, agreeableness, and conscientiousness.

**Bishop:** a chess piece that may move to any square along a diagonal on which it stands (three points).

**Blindfold chess:** a modality of chess playing in which looking at the board is disallowed. The player is verbally informed about the moves in order to be able to memorize them and keep track of the development of the game.

**Blitz:** a chess game with a severe constraint in the available thinking time. Each player's thinking time is ten minutes or less.

**Brown–Peterson paradigm:** a cognitive experiment designed to test the boundaries of working memory. It evaluates working memory while avoiding using memorization techniques to increase recall by interpolating a distracting task.

**Castling:** a combined chess move of the king and either rook of the same colour along the player's first rank. This move is mainly aimed at safeguarding the king.

**Centipede game:** an econometric finite-move two-player game whereby both players choose alternately to finish or to pass to the other player. The payoff received by one player when ending the game is greater than the payoff when the other player ends the game, but lower when the other player continues with the game.

**Check:** a chess position that arises when the king is placed under direct attack by one or more pieces from the opponent. There are three possibilities for the king when in the check state: (1) moving to a different empty square uncontrolled by the opponent's pieces; (2) blocking the check with another piece; (3) capturing the attacking piece. When none of these three possibilities is feasible, there is a checkmate position.

**Checkmate:** the end position arising in a chess game whereby the opponent's king is placed under attack (check) in a way that there is no legal move to avoid the check. When player A checkmates player B, it means that player B's king has been checkmated and loses the game, and player A wins the game.

**Chess tree:** a conceptual device that is useful to store and keep track of all the possible combinations of moves within a chess game.

**Chunk:** a grouping of four, five, or more chess pieces that bear a relational structure between them, and that can be stored in memory.

**Correlation:** the degree of association between two quantitative variables, which varies between minus one and one. Values closer to zero indicate a lower association between both variables. Higher departures from zero, either negative or positive, indicate a higher association between both variables.

**Crystallized intelligence:** the abilities tapped by tests dealing with general knowledge, use of language, and learned skills. Crystallized abilities are thought to underlie the environmental component of intelligence.

**Default brain network:** the group of interconnected brain structures that are active in a resting state.

**Draw:** the outcome of a chess game in which each opponent wins half a point and neither of the two players wins or loses.

**EEG (Electroencephalography):** the bioelectrical signal that characterizes the electrical activity of the brain.

**Einstellung effect:** the predisposition to solving a problem in a certain way despite the fact that there might be more appropriate methods.

**Elo rating:** a chess rating system devised by Arpad Elo (Elo, 1978), used to quantify chess players' skill. In general, higher Elo ratings denote a higher level of chess skill.

**Endgame:** the final part of a chess game, in which there is a notably smaller amount of playing pieces.

**Factor analysis:** a statistical technique used in psychology to summarize data generally obtained with a psychological test from several individuals. It permits the obtaining of a few unobservable or latent factors that have psychological meaning and can be interpreted for different purposes.

**FIDE (Fédération Internationale des Échecs):** International World Chess Federation.

**Flicker paradigm:** a perceptual experiment whereby image I is repeatedly and quickly switched back and forth with a slightly changed image I'. The subject is instructed to indicate as soon as possible a detected change between I and I'.

**Fluid intelligence:** the abilities tapped by tests dealing with the resolution of novel and abstract problems quickly and efficiently. Fluid abilities are thought to underlie the genetic component of intelligence.

**Gambit:** a chess opening in which a player sacrifices a piece, most often a pawn, in exchange for some kind of strategic or tactical advantage.

**Game theory:** the study of a wide assortment of mathematical models intended to represent the interaction of two or more decision-makers.

**General intelligence:** also known as the *g* factor. A very broad ability measured by cognitive abilities tests, which can be derived by the factor analysis technique.

**Genotype:** the assortment of an organism's genes that determine its characteristics.

**Grandmaster (GM):** a chess title issued to very strong players by the FIDE. A grandmaster usually has over 2500 Elo points.

**Heritability:** the proportion of phenotypic variance due to genetic factors.

**Heuristic:** an approach to problem solving based on accumulated experience, whereby the search in a massive solution space is encouraged to find an optimal solution.

**Individual differences:** the observable variation in psychological traits measured by psychometric tests and by other instruments and methods.

**Intelligence quotient (IQ):** a quantitative measure of human intelligence characterized as the ratio of mental age (MA) to chronological age (CA): IQ = [MA / CA]*100. Typically, the IQ has a mean of 100 and a standard deviation of fifteen.

**International master (IM):** the second-highest title in chess; it precedes the grandmaster (GM) title.

**King:** a chess piece that may move to any adjoining square or by castling.

**Kingside:** the side of the chessboard comprising columns e, f, g, and h.

**Knight:** a chess piece that may move to one of the squares nearest to that on which it stands, but not on the same rank, file, or diagonal. It is the only piece that can jump over the other pieces on the board, which makes it particularly tricky (three points).

**Knight's row task:** a chess test used to estimate chess skill derived from the knight's tour, a mathematical problem that consists of finding a route for the knight piece through the chessboard. The piece has to land on each square only once until all squares have been visited. It corresponds to a Hamiltonian graph in graph theory.

**Lability (emotional):** changes in mood that are highly pronounced.

**Meta-analysis:** a quantitative review of single empirical studies sharing a specific topic that attempts to summarize or shed further light on this topic or on a particular hypothesis.

**Middlegame:** the central part of a chess game, which occurs after the opening and before the endgame, though this separation is ambiguous.

**Neural efficiency hypothesis (NEH):** a model about human intelligence. It predicts that, when facing an intellectually demanding task, the brain activation required to solve the task should be lower for people with higher intelligence and higher for people with lower intelligence.

**Opening:** the initial moves in a chess game, characterized by the early development of the pieces, control of the central part of the board (squares e4, d4, e5, d5), and the safety of the king.

**Orthogonal factor solution:** a solution derived from a factor analysis, in which the obtained factors are highly independent and uncorrelated. An oblique factor solution, in contrast, yields factors that are highly correlated.

**Pawn:** a chess piece that may move to the square in front of it on the same file if this square is unoccupied, or two squares from the initial position, or to a square occupied by an opponent's piece diagonally in front of it on an adjacent file, capturing that piece (one point).

**Phenotype:** the morphological, developmental, and behavioural observable characteristics in an organism.

**Ply:** one turn taken by one player. A chess move comprises two plies: one ply for white, and one ply for black.

**Power law of practice:** a mathematical function that has been used to predict performance from practice. Practice improves performance quickly to a certain point. Further improvements beyond that point due to practice tend to be gradually lower.

**PPIK:** a psychological theory, developed by Philip Ackerman, that can be used to explain the structure and development of chess expertise. It is composed of four main broad groups of traits: intelligence as process, personality, interests, and intelligence as knowledge.

**Psychometrics:** a field of research within psychology that aims to devise methods and instruments intended to characterize individual differences in psychological traits.

**Queen:** a chess piece that may move to any square along the file, the rank, or a diagonal on which it stands (nine points).

**Queenside:** the side of the chessboard comprising columns a, b, c, and d.

**Race to 100 game:** an econometric two-player game whereby players take turns choosing numbers between one and ten. A sum of all the chosen numbers is kept, and the first player who makes the total equal to 100 wins the game.

**Relative age effect (RAE):** the performance gap due to maturation differences within a group of young people selected on the basis of a period, usually a year. The RAE tends to favour older over younger individuals.

**Rook:** a chess piece that may move to any square along the file or the rank on which it stands (four points).

**Saccade:** a simultaneous movement of both eyes in the same direction.

**Simultaneous chess:** a chess competition in which one player, generally at the grandmaster level, plays against more than one opponent at the same time.

**Theory of mind (ToM):** the ability to understand and predict the mental state of others.

**Tower of London test:** a neuropsychological test to evaluate deficits in planning abilities.

**Trait:** an observable psychological characteristic that is variable between and within individuals.

**Universal Rating System (URS):** an alternative chess rating system to the commonly accepted Elo rating. It takes into account performance in both slow and fast chess games.

**Zugzwang:** a chess circumstance whereby any move made by one player will worsen his or her position.

# REFERENCES


Abramson, L. Y., Seligman, M. E. P., and Teasdale, J. D. (1978). Learned helplessness in humans: critique and reformulation. *Journal of Abnormal Psychology, 87* (1), 49–74.

Acharya, J. N., Hani, A., Cheek, J., Thirumala, P., and Tsuchida, T. N. (2016). American Clinical Neurophysiology Society Guideline 2: guidelines for standard electrode position nomenclature. *Journal of Clinical Neurophysiology, 33*(4), 308–311.

Achten, J., and Jeukendrup, A. E. (2003). Heart rate monitoring: applications and limitations. *Sports Medicine, 33*(7), 517–538.

Aciego, R., García, L., and Betancort, M. (2012). The benefits of chess for the intellectual and social-emotional enrichment in schoolchildren. *Spanish Journal of Psychology, 15*(2), 551–559.

Aciego, R., García, L., and Betancort, M. (2016). Efectos del método de entrenamiento en ajedrez, entrenamiento táctico versus formación integral, en las competencias cognitivas y sociopersonales de los escolares. *Universitas Psychologica, 15*(1), 165–176.

Ackerman, P. L. (1987). Individual differences in skill learning: an integration of psychometric and information processing perspectives. *Psychological Bulletin, 102*(1), 3–27.

Ackerman, P. L. (1996). A theory of adult intellectual development: process, personality, interests, and knowledge. *Intelligence, 22*(2), 227–257.

Ackerman, P. L. (2007). New developments in understanding skilled performance. *Current Directions in Psychological Science, 16*(5), 235–239.

Ackerman, P. L. (2011). Intelligence and expertise. In R. J. Sternberg and S. B. Kaufman (eds.), *The Cambridge Handbook of Intelligence*. New York: Cambridge University Press, 847–860.

Ackerman, P. L. (2014). Nonsense, common sense, and science of expert performance: talent and individual differences. *Intelligence, 45*(1), 6–17.

Ackerman, P. L., and Heggestad, E. D. (1997). Intelligence, personality, and interests: evidence for overlapping traits. *Psychological Bulletin, 121*(2), 219–245.

Albers, P. C. H., and de Vries, H. (2001). Elo-rating as a tool in the sequential estimation of dominance strengths. *Animal Behaviour, 61*, 489–495.

Allport, G. W., and Odbert, H. S. (1936). Trait names: a psycholexical study. *Psychological Monographs, 47*(1), i–171.

Aluja, A., and Blanch, A. (2004). Socialized personality, scholastic aptitudes, study habits, and academic achievement: exploring the link. *European Journal of Psychological Assessment, 20*(3), 157–165.

Amidzic, O., Riehle, H. J., and Elbert, T. (2006). Toward a psychophysiology of expertise: focal magnetic gamma bursts as a signature of memory chunks and the aptitude of chess players. *Journal of Psychophysiology, 64*(4), 370–381.

Amidzic, O., Riehle, H. J., Fehr, T., Wienbruch, C., and Elbert, T. (2001). Pattern of focal gamma-bursts in chess players. *Nature, 412*, 603.

Amunts, K., and Zilles, K. (2015). Architectonic mapping of the human brain beyond Brodmann. *Neuron, 88*(6), 1086–1107.

Anastasi, A. (1937). *Differential Psychology: Individual and Group Differences in Behavior*. New York: Macmillan.

Anastasi, A. (1958). Heredity, environment, and the question 'how?'. *Psychological Review, 65*(4), 197–208.

Anderson, C. A., and Bushman, B. J. (2002). Human aggression. *Annual Review of Psychology, 53*, 27–51.

Anderson, M. (2010). Neural reuse: a fundamental organizational principle of the brain. *Behavioral and Brain Sciences, 33*(4), 245–313.

Andrés-Pueyo, A. (1997). *Manual de psicología diferencial*. Madrid: McGraw-Hill.

Antal, M. (2013). On the use of ELO rating for adaptive assessment. *Studia Universitatis Babes-Bolyai, 58*(1), 29–41.

Ashburner, M., Ball, C. A., Blake, J. A., Botstein, D., Butler, H., Cherry, J. M., et al. (2000). Gene ontology: tool for the unification of biology. *Nature Genetics, 25*(1), 25–29.

Atherton, M., Zhuang, J., Bart, W. M., Hu, X., and He, S. (2003). A functional MRI study of high-level cognition, I: the game of chess. *Cognitive Brain Research, 16* (1), 26–31.

Avni, A., Kipper, D. A., and Fox, S. (1987). Personality and leisure activities: an illustration with chess players. *Personality and Individual Differences, 8*(5), 715–719.

Aydin, M. (2015). Examining the impact of chess instruction for the visual impairment on mathematics. *Educational Research and Reviews, 10*(7), 907–911.

Bachmann, T., and Oit, M. (1992). Stroop-like interference in chess players' imagery: an unexplored possibility to be revealed by the adapted moving-spot task. *Psychological Research, 54*(1), 27–31.

Baddeley, A., Thomson, N., and Buchanan, M. (1975). Word length and the structure of short-term memory. *Journal of Verbal Learning and Verbal Behavior, 14*(6), 575–589.

Baker, J., and Young, B. (2014). 20 years later: deliberate practice and the development of expertise in sport. *International Review of Sport and Exercise Psychology, 7*(1), 135–157.

Baltes, P. B. (1987). Theoretical propositions of life-span developmental psychology: on the dynamics between growth and decline. *Developmental Psychology, 23*(5), 611–626.

Barrett, D. C., and Fish, W. W. (2011). Our move: using chess to improve math achievement for students who receive special education services. *International Journal of Special Education, 26*(3), 181–193.

Barrick, M. R., and Mount, M. K. (1991). The big five personality dimensions and job performance: a meta-analysis. *Personnel Psychology, 44*(1), 1–26.

Bart, W. M. (2014). On the effect of chess training on scholastic achievement. *Frontiers in Psychology, 5*, article 762.

Batchelder, W. H., and Bershad, N. J. (1979). The statistical analysis of a Thurstonian model for rating chess players. *Journal of Mathematical Psychology, 19*(1), 39–60.

Belsky, D. W., Moffitt, T. E., Corcoran, D. L., Domingue, B., Harrington, H., Hogan, S., et al. (2016). The genetics of success: how single-nucleotide polymorphisms associated with educational attainment relate to life-course development. *Psychological Science, 27*(7), 957–972.

Benbow, C. P. (1988). Sex differences in mathematical reasoning ability among the intellectually talented: their nature, effects, and possible causes. *Behavioral and Brain Sciences, 11*(2), 169–232.

Benbow, C. P., Lubinski, D., Shea, D. L., and Eftekhari-Sanjani, H. (2000). Sex differences in mathematical reasoning ability at age 13: their status 20 years later. *Psychological Science, 11*(6), 474–480.

Benbow, C. P., and Stanley, J. C. (1980). Sex differences in mathematical ability: fact or artifact? *Science, 210*, 1262–1264.

Benko, P., and Hochberg, B. (1991). *Winning with Chess Psychology*. New York: Random House.

Bennett, G. K., Seashore, H. G., and Wesman, A. G. (2002). *Differential Aptitude Tests*, 6th edn. San Antonio, TX: Pearson.

Berliner, H. J. (1974). Chess as problem solving: the development of a tactics analyzer. Unpublished doctoral thesis, Carnegie Mellon University, Pittsburgh.

Berthelot, G., Len, S., Hellard, P., Tafflet, M., Guillaume, M., Vollmer, J. C., et al. (2012). Exponential growth combined with exponential decline explains lifetime performance evolution in individual and human species. *Age, 34*(4), 1001–1009.

Bertoni, M., Brunello, G., and Rocco, L. (2015). Selection and the age–productivity profile: evidence from chess players. *Journal of Economic Behavior and Organization, 110*, 45–58.

Betsch, T., and Glöckner, A. (2010). Intuition in judgment and decision making: extensive thinking without effort. *Psychological Inquiry, 21*(4), 279–294.

Bettencourt, B. A., Talley, A., Benjamin, A. J., and Valentine, J. (2006). Personality and aggressive behavior under provoking and neutral conditions: a meta-analytic review. *Psychological Bulletin, 132*(5), 751–777.

Beus, J. M., Dhanani, L. Y., and McCord, M. A. (2015). A meta-analysis of personality and workplace safety: addressing unanswered questions. *Journal of Applied Psychology, 100*(2), 481–498.

Bilalić, M. (2017). *The Neuroscience of Expertise.* Cambridge: Cambridge University Press.

Bilalić, M. (2018). The double take of expertise: neural expansion is associated with outstanding performance. *Current Directions in Psychological Science, 27*(6), 462–469.

Bilalić, M., Kiesel, A., Pohl, C., Erb, M., and Grodd, W. (2011). It takes two: skilled recognition of objects engages lateral areas in both hemispheres. *PLOS ONE, 6* (1), e16202.

Bilalić, M., Langner, R., Erb, M., and Grodd, W. (2010). Mechanisms and neural basis of object and pattern recognition: a study with chess experts. *Journal of Experimental Psychology: General, 139*(4), 728–742.

Bilalić, M., Langner, R., Ulrich, R., and Grodd, W. (2011). Many faces of expertise: fusiform face area in chess experts and novices. *Journal of Neuroscience, 31*(28), 10206–10214.

Bilalić, M., and McLeod, P. (2006). How intellectual is chess? A reply to Howard. *Journal of Biosocial Science, 38*(3), 419–421.

Bilalić, M., and McLeod, P. (2007). Participation rates and the differences in performance of women and men in chess. *Journal of Biosocial Science, 39*(5), 789–793.

Bilalić, M., McLeod, P., and Gobet, F. (2007a). Does chess need intelligence? A study with young chess players. *Intelligence, 35*(5), 457–470.

Bilalić, M., McLeod, P., and Gobet, F. (2007b). Personality profiles of young chess players. *Personality and Individual Differences, 42*(6), 901–910.

Bilalić, M., McLeod, P., and Gobet, F. (2008a). Inflexibility of experts: reality or myth? Quantifying the Einstellung effect in chess masters. *Cognitive Psychology, 56*(2), 73–102.

Bilalić, M., McLeod, P., and Gobet, F. (2008b). Why good thoughts block better ones: the mechanism of the pernicious Einstellung (set) effect. *Cognition, 108*(3), 652–661.

Bilalić, M., McLeod, P., and Gobet, F. (2009). Specialization effect and its influence on memory and problem solving in expert chess players. *Cognitive Science, 33* (6), 1117–1143.

Bilalić, M., Smallbone, K., McLeod, P., and Gobet, F. (2009). Why are (the best) women so good at chess? Participation rates and gender differences in intellectual domains. *Proceedings of the Royal Society B, 276*, 1161–1165.

Bilalić, M., Turella, L., Campitelli, G., Erb, M., and Grodd, W. (2012). Expertise modulates the neural basis of context-dependent recognition of objects and their relations. *Human Brain Mapping, 33*(11), 2728–2740.

Binet, A. (1893). Les grandes mémoires: résume d'une enquête sur les joueurs d'échecs [Mnemonic virtuosity: a study of chess players]. *Revue des deux mondes, 117*, 826–859.

Binet, A. (1894). *Psychologie des grands calculateurs et des joueurs d'échecs*. Paris: Hachette.

Blanch, A. (2016). Expert performance of men and women: a cross-cultural study in the chess domain. *Personality and Individual Differences, 101*, 90–97.

Blanch, A. (2018). Top hundred chess experts: a cross-domain analysis of change over time. *Intelligence, 71*, 76–84.

Blanch, A., and Aluja, A. (2013). A regression tree of the aptitudes, personality, and academic performance relationship. *Personality and Individual Differences, 54* (6), 703–708.

Blanch, A., Aluja, A., and Cornadó, M. P. (2015). Sex differences in chess performance: analyzing participation rates, age, and practice in chess tournaments. *Personality and Individual Differences, 86*, 117–121.

Blanch, A., García, H., Llaveria, A., and Aluja, A. (2017). The Spearman's law of diminishing returns in chess. *Personality and Individual Differences, 104*, 434–441.

Blanch, A., García, R., Planes, J., Gil, R., Balada, F., Blanco, E., et al. (2017). Ontologies about human behavior: a review of knowledge modeling systems. *European Psychologist, 22*(3), 180–197.

Blasco-Fontecilla, H., Gonzalez-Perez, M., Garcia-Lopez, R., Poza-Cano, B., Perez-Moreno, M. R., de Leon-Martinez, V., et al. (2016). Eficacia del ajedrez en el tratamiento del trastorno por déficit de atención e hiperactividad: un estudio prospectivo abierto. *Revista de Psiquiatría y Salud Mental, 9*(1), 13–21.

Bleidorn, W., Kandler, C., Hülsheger, U. R., Riemann, R., Angleitner, A., and Spinath, F. M. (2010). Nature and nurture of the interplay between personality traits and major life goals. *Journal of Personality and Social Psychology, 99*(2), 366–379.

Boggan, A. L., Bartlett, J. C., and Krawczyk, D. C. (2012). Chess masters show a hallmark of face processing with chess. *Journal of Experimental Psychology: General, 141*(1), 37–42.

Bollen, K. A. (1989). *Structural Equations with Latent Variables*. New York: Wiley.

Borkenau, P., McCrae, R. R., and Terracciano, A. (2013). Do men vary more than women in personality? A study in 51 cultures. *Journal of Research in Personality, 47*(2), 135–144.

Bouchard, T. J., and McGue, M. (1981). Familial studies of intelligence: a review. *Science, 212*, 1055–1059.

Boyle, G. J., Stankov, L., and Cattell, R. B. (1995). Measurement and statistical models in the study of personality and intelligence. In M. Zeidner (ed.), *International Handbook of Personality and Intelligence*. New York: Plenum, 417–446.

Bradley, R. A., and Terry, M. E. (1952). Rank analysis of incomplete block designs, I: the method of paired comparisons. *Biometrika, 39*(3/4), 324–345.

Breznik, K., and Batagelj, V. (2011). FIDE chess network. *Austrian Journal of Statistics, 40*(4), 225–239.

Breznik, K., and Law, K. M. Y. (2016). Relative age effect in mind games: the evidence from elite chess. *Perceptual and Motor Skills*, *122*(2), 583–594.

Briley, D. A., and Tucker-Drob, E. M. (2014). Genetic and environmental continuity in personality development: a meta-analysis. *Psychological Bulletin*, *140*(5), 1303–1331.

Bronfenbrenner, U., and Ceci, S. J. (1994). Nature–nurture reconceptualized in developmental perspective: a bioecological model. *Psychological Review*, *101*(4), 568–586.

Browne, K. R. (2006). Evolved sex differences and occupational segregation. *Journal of Organizational Behavior*, *27*(2), 143–162.

Bühren, C., and Frank, B. (2012). Chess players' performance beyond 64 squares: a case study on the limitations of cognitive abilities transfer. *Talent Development and Excellence*, *4*(2), 157–169.

Bukach, C. M., Gauthier, I., and Tarr, M. J. (2006). Beyond faces and modularity: the power of an expertise framework. *Trends in Cognitive Sciences*, *10*(4), 159–166.

Burgoyne, A. P., Sala, G., Gobet, F., Macnamara, B. N., Campitelli, G., and Hambrick, D. Z. (2016). The relationship between cognitive ability and chess skill: a comprehensive meta-analysis. *Intelligence*, *59*, 72–83.

Burns, B. D. (2004). The effects of speed on skilled chess performance. *Psychological Science*, *15*(7), 442–447.

Buss, D. M. (1991). Evolutionary personality psychology. *Annual Review of Psychology*, *42*, 459–491.

Buss, D. M. (2001). Human nature and culture: an evolutionary psychological perspective. *Journal of Personality*, *69*(6), 955–978.

Buss, D. M. (2005). *The Handbook of Evolutionary Psychology*. Hoboken, NJ: Wiley.

Buss, D. M. (2009). How can evolutionary psychology successfully explain personality and individual differences? *Perspectives on Psychological Science*, *4*(4), 359–366.

Calderwood, R., Klein, G., and Crandall, B. (1988). Time pressure, skill, and move quality in chess. *American Journal of Psychology*, *101*(4), 481–493.

Cameron, C. E., Grimm, K. J., Steele, J. S., Castro-Schilo, L., and Grissmer, D. W. (2015). Nonlinear Gompertz curve models of achievement gaps in mathematics and reading. *Journal of Educational Psychology*, *107*(3), 789–804.

Campbell, M., Hoane, A. J., and Hsu, F. (2002). Deep Blue. *Artificial Intelligence*, *134*, 57–83.

Campitelli, G., and Gobet, F. (2004). Adaptive expert decision making: skilled chess players search more and deeper. *Journal of International Computer Games Association*, *27*(4), 209–216.

Campitelli, G., and Gobet, F. (2005). The mind's eye in blindfold chess. *European Journal of Cognitive Psychology*, *17*(1), 23–45.

Campitelli, G., and Gobet, F. (2008). The role of practice in chess: a longitudinal study. *Learning and Individual Differences*, *18*(4), 446–458.

Campitelli, G., and Gobet, F. (2011). Deliberate practice: necessary but not sufficient. *Current Directions in Psychological Science, 20*(5), 280–285.

Campitelli, G., Gobet, F., Head, K., Buckley, M., and Parker, A. (2007). Brain localization of memory chunks in chess players. *International Journal of Neuroscience, 117*(12), 1641–1659.

Campitelli, G., Gobet, F., and Parker, A. (2005). Structure and stimulus familiarity: a study of memory in chess-players with functional magnetic resonance imaging. *Spanish Journal of Psychology, 8*(2), 238–245.

Campitelli, G., Parker, A., Head, K., and Gobet, F. (2008). Left lateralization in autobiographical memory: an fMRI study using the expert archival paradigm. *International Journal of Neuroscience, 118*(2), 191–209.

Cannice, M. V. (2013). The right moves: creating experiential management learning with chess. *International Journal of Management Education, 11*(1), 25–33.

Carretié, L. (2001). *Psicofisiologia*. Madrid: Ediciones Pirámide.

Carroll, J. B. (1993). *Human Cognitive Abilities: A Survey of Factor-Analysis Studies*. Cambridge: Cambridge University Press.

Caspi, A., and Herbener, E. S. (1990). Continuity and change: assortative marriage and the consistency of personality in adulthood. *Journal of Personality and Social Psychology, 58*(2), 250–258.

Caspi, A., Roberts, B. W., and Shiner, R. L. (2005). Personality development: stability and change. *Annual Review of Psychology, 56*, 453–484.

Cattell, R. B. (1963). Theory of fluid and crystallized intelligence: a critical experiment. *Journal of Educational Psychology, 54*(1), 1–22.

Cattell, R. B. (1987). *Intelligence: Its Structure, Growth and Action*. Amsterdam: North-Holland.

Cattell, R. B., Eber, H. W., and Tatsuoka, M. M. (1970). *Handbook for the Sixteen Personality Factor Questionnaire (16PF)*. Champaign, IL: Institute for Personality and Ability Testing.

Cattell, R. B., and Kline, P. (1977). *The Scientific Analysis of Personality and Motivation*. London: Academic Press.

Ceci, S. J., and Williams, W. M. (2011). Understanding current causes of women's underrepresentation in science. *Proceedings of the National Academy of Sciences of the United States of America, 108*(8), 3157–3162.

Ceci, S. J., Williams, W. M., and Barnett, S. M. (2009). Women's underrepresentation in science: sociocultural and biological considerations. *Psychological Bulletin, 135*(2), 218–261.

Chabris, C. F. (2017). Six suggestions for research on games in cognitive science. *Topics in Cognitive Science, 9*(2), 497–509.

Chabris, C. F., and Glickman, E. M. (2006). Sex differences in intellectual performance: analysis of a large cohort of competitive chess players. *Psychological Science, 17*(12), 1040–1046.

Chabris, C. F., and Hamilton, S. E. (1992). Hemispheric specialization for skilled perceptual organization by chessmasters. *Neuropsychologia, 30*(1), 47–57.

Chabris, C. F., and Hearst, E. S. (2003). Visualization, pattern recognition, and forward search: effects of playing speed and sight of the position on grandmaster chess errors. *Cognitive Science, 27*(4), 637–648.

Chabris, C. F., Hebert, B. M., Benjamin, D. J., Beauchamp, J. P., Cesarini, D., van der Loos, M. J. H. M., et al. (2012). Most reported genetic associations with general intelligence are probably false positives. *Psychological Science, 20*(5), 1–10.

Chabris, C. F., Lee, J. J., Benjamin, D. J., Beauchamp, J. P., Glaeser, E. L., Borst, G., et al. (2013). Why it is hard to find genes associated with social science traits: theoretical and empirical considerations. *American Journal of Public Health, 103* (S1), S152–S166.

Chang, Y. A., and Lane, D. M. (2016). There is time for calculation in speed chess and calculation accuracy increases with expertise. *American Journal of Psychology, 129*(1), 1–9.

Charness, N. (1976). Memory for chess positions: resistance to interference. *Journal of Experimental Psychology: Human Learning and Memory, 2*(6), 641–653.

Charness, N. (1981). Aging and skilled problem solving. *Journal of Experimental Psychology: General, 110*(1), 21–38.

Charness, N. (1992). The impact of chess research on cognitive science. *Psychological Research, 54*(1), 4–9.

Charness, N., and Gerchak, Y. (1996). Participation rates and maximal performance: a log-linear explanation for group differences, such as Russian and male dominance in chess. *Psychological Science, 7*(1), 46–51.

Charness, N., Reingold, E., Pomplun, M., and Stampe, D. (2001). The perceptual aspect of skilled performance in chess: evidence from eye movements. *Memory and Cognition, 29*(8), 1146–1152.

Charness, N., Tuffiash, M., Krampe, R., Reingold, E., and Vasyukova, E. (2005). The role of deliberate practice in chess expertise. *Applied Cognitive Psychology, 19*(2), 151–165.

Chase, W. G., and Simon, H. A. (1973). Perception in chess. *Cognitive Psychology, 4* (1), 55–81.

Chassy, P., and Gobet, F. (2011). Measuring chess experts' single-use sequence knowledge: an archival study of departure from 'theoretical' openings. *PLOS ONE, 6*(11), e26692.

Chi, M. T. H. (1978). Knowledge structures and memory development. In R. S. Siegler (ed.), *Children's Thinking: What Develops?* Hillsdale, NJ: Lawrence Erlbaum Associates, 73–96.

Chi, M. T. H. (2006a). Two approaches to the study of experts' characteristics. In K. A. Ericsson, N. Charness, P. J. Feltovich, and R. R. Hoffman (eds.), *The Cambridge Handbook of Expertise and Expert Performance.* New York: Cambridge University Press, 21–30.

Chi, M. T. H. (2006b). Laboratory methods for assessing experts' and novices' knowledge. In K. A. Ericsson, N. Charness, P. J. Feltovich, and R. R. Hoffman (eds.), *The Cambridge Handbook of Expertise and Expert Performance*. New York: Cambridge University Press, 167–184.

Christiaen, J., and Verhofstadt-Denève, L. (1981). Schaken en cognitieve ontwikkeling. *Nederlands Tijdschrift voor de Psychologie en haar Grensgebieden*, *36*, 561–582.

Church, A. T. (2001). Personality measurement in cross-cultural perspective. *Journal of Personality*, *69*(6), 979–1006.

Church, A. T. (2010). Current perspectives in the study of personality across cultures. *Perspectives on Psychological Science*, *5*(4), 441–449.

Clark, D. L., Boutros, N. N., and Mendez, M. F. (2010). *The Brain and Behavior: An Introduction to Behavioral Neuroanatomy*. New York: Cambridge University Press.

Cleveland, A. (1907). The psychology of chess and of learning to play it. *American Journal of Psychology*, *18*(3), 269–308.

Cohen, J. (1988). *Statistical Power Analysis for the Behavioral Sciences*. Hillsdale, NJ: Lawrence Erlbaum Associates.

Colom, R. (2005). *Psicología de las diferencias individuales: Teoría y práctica*. Madrid: Ediciones Pirámide.

Colom, R., Haier, R. J., Head, K., Álvarez-Linera, J., Quiroga, M. A., Chun Shih, P., et al. (2009). Gray matter correlates of fluid, crystallized, and spatial intelligence: testing the P-FIT model. *Intelligence*, *37*(2), 124–135.

Conley, J. J. (1984). The hierarchy of consistency: a review and model of longitudinal findings on adult individual differences in intelligence, personality and self-opinion. *Personality and Individual Differences*, *5*(1), 11–25.

Connors, M. H., Burns, B. D., and Campitelli, G. (2011). Expertise in complex decision making: the role of search in chess 70 years after de Groot. *Cognitive Science*, *35*(8), 1567–1579.

Cooke, N., Atlas, R., Lane, D., and Berger, R. (1993). Role of high-level knowledge in memory for chess positions. *American Journal of Psychology*, *106*(3), 321–351.

Corr, P. J. (2004). Reinforcement sensitivity theory and personality. *Neuroscience and Biobehavioral Reviews*, *28*(3), 317–332.

Costa, P. T., and McCrae, R. R. (1985). *The NEO Personality Inventory*. Odessa, FL: Psychological Assessment Resources.

Costa, P. T., and McCrae, R. R. (1992). *Professional Manual: Revised NEO Personality Inventory (NEO-PI-R) and NEO Five-Factor Inventory (NEO-FFI)*. Odessa, FL: Psychological Assessment Resources.

Crawford, C., Dearden, L., and Meghir, C. (2007). *When You Are Born Matters: The Impact of Date of Birth on Child Cognitive Outcomes in England*. London: Institute for Fiscal Studies.

Cronbach, L. J. (1957). The two disciplines of scientific psychology. *American Psychologist*, *12*(11), 671–684.

Cronbach, L. J., and Webb, N. (1975). Between-class and within-class effects in a reported aptitude X treatment interaction: reanalysis of a study by G. L. Anderson. *Journal of Educational Psychology*, *67*(6), 717–724.

Das, J. P. (1999). A neo-Lurian approach to assessment and remediation. *Neuropsychology Review*, *9*(2), 107–116.

Davidson, J. E., and Kemp, I. A. (2011). Contemporary models of intelligence. In R. J. Sternberg and S. B. Kaufman (eds.), *The Cambridge Handbook of Intelligence*. New York: Cambridge University Press, 58–78.

Davidson, R. J., Putnam, K. M., and Larson, C. L. (2000). Dysfunction in the neural circuitry of emotion regulation: a possible prelude to violence. *Science*, *289*, 591–594.

Davis, K. L., and Panksepp, J. (2011). The brain's emotional foundations of human personality and the Affective Neuroscience Personality Scales. *Neuroscience and Biobehavioral Reviews*, *35*(9), 1946–1958.

De Bolle, M., De Fruyt, F., McCrae, R. R., Löckenhoff, C. E., Costa, P. T., Aguilar-Vafaie, M. E., et al. (2015). The emergence of sex differences in personality traits in early adolescence: a cross-sectional, cross-cultural study. *Journal of Personality and Social Psychology*, *108*(1), 178–185.

De Bruin, A. B., Kok, E., Leppink, J., and Camp, G. (2014). Practice, intelligence, and enjoyment in novice chess players: a prospective study at the earliest stage of a chess career. *Intelligence*, *45*, 18–25.

De Bruin, A. B., Rikers, R. M. J. P., and Schmidt, H. G. (2005). Monitoring accuracy and self-regulation when learning to play a chess endgame. *Applied Cognitive Psychology*, *19*(2), 167–181.

De Bruin, A. B., Rikers, R. M. J. P., and Schmidt, H. G. (2007a). The effect of self-explanation and prediction on the development of principled understanding of chess in novices. *Contemporary Educational Psychology*, *32*(2), 188–205.

De Bruin, A. B., Rikers, R. M. J. P., and Schmidt, H. G. (2007b). The influence of achievement motivation and chess-specific motivation on deliberate practice. *Journal of Sport and Exercise Psychology*, *29*(5), 561–583.

De Bruin, A. B., Smits, N., Rikers, R., and Schmidt, H. (2008). Deliberate practice predicts performance over time in adolescent chess players and drop-outs: a linear mixed models analysis. *British Journal of Psychology*, *99* (4), 473–497.

De Groot, A. D. (1965). *Thought and Choice in Chess*, 2nd edn. The Hague: Mouton.

De Groot, A. D., Gobet, F., and Jongman, R. W. (1996). *Perception and Memory in Chess: Studies in the Heuristics of the Professional Eye*. Assen, Netherlands: Van Gorcum.

Deary, I. J. (2001). Human intelligence differences: towards a combined experimental-differential approach. *Trends in Cognitive Science*, *5*(4), 164–170.

Deary, I. J., and Stough, C. (1996). Intelligence and inspection time: achievements, prospects and problems. *American Psychologist, 51*(6), 599–608.

Deary, I. J., Strand, S., Smith, P., and Fernandes, C. (2007). Intelligence and educational achievement. *Intelligence, 35*(1), 13–21.

Deary, I. J., Whalley, L. J., Lemmon, H., Crawford, J. R., and Starr, J. M. (2000). The stability of individual differences in mental ability from childhood to old age: follow-up of the 1932 Scottish mental survey. *Intelligence, 28*(1), 49–55.

Demily, C., Desmurget, M., Chambon, V., and Franck, N. (2009). The game of chess enhances cognitive abilities in schizophrenia. *Schizophrenia Research, 107* (1), 112–113.

DeNeve, K. M., and Cooper, H. (1998). The happy personality: a meta-analysis of 137 personality traits and subjective well-being. *Psychological Bulletin, 124*(2), 197–229.

Di Fatta, G., McHaworth, G., and Regan, K. W. (2009). Skill rating by Bayesian inference. Paper presented at the IEEE Symposium on Computational Intelligence and Data Mining, Nashville, 30 March.

Dietrich, A., and Kanso, R. (2010). A review of EEG, ERP, and neuroimaging studies of creativity and insight. *Psychological Bulletin, 136*(5), 822–848.

Djakow, I. N., Petrowski, N. W., and Rudik, P. A. (1927). *Psychologie des Schachspiels*. Berlin: de Gruyter.

Doll, J., and Mayr, U. (1987). Intelligenz und Schachleistung: eine Untersuchung an Schachexperten. *Psychologische Beiträge, 29*(2/3), 270–289.

Doppelmayr, M., Klimesch, W., Stadler, W., Pöllhuber, D., and Heine, C. (2002). EEG alpha power and intelligence. *Intelligence, 30*(3), 289–302.

Draper, N. R. (1963). Does age affect master chess? *Journal of the Royal Statistical Society, 126*(1), 120–127.

Dreber, A., Gerdes, C., and Gränsmark, P. (2013). Beauty queens and battling knights: risk taking and attractiveness in chess. *Journal of Economic Behavior and Organization, 90*(1), 1–18.

Duan, X., He, S., Liao, W., Liang, D., Qiu, L., Wei, L., et al. (2012). Reduced caudate volume and enhanced striatal–DMN integration in chess experts. *NeuroImage, 60*(2), 1280–1286.

DuCette, J. (2009). An evaluation of the Chess Challenge Program of ASAP/After School Activities Partnerships. Philadelphia: After School Activities Partnerships.

Dunn L. M., and Dunn, D. M. (1997). *Peabody Picture Vocabulary Test*, 4th edn. San Antonio, TX: Pearson.

Duru, A. D., and Assem, M. (2018). Investigating neural efficiency of elite karate athletes during a mental arithmetic task using EEG. *Cognitive Neurodynamics, 12*(1), 95–102.

Dvoretsky, M., and Yusupov, A. (1996). *Positional Play*. London: Batsford.

Eberhard, J. W. (2003). The relationship between classroom chess instruction and verbal, quantitative, and nonverbal reasoning abilities of economically disadvantaged students. Unpublished doctoral dissertation, Texas A&M University, College Station.

Elo, A. (1978). *The Rating of Chessplayers, Past and Present*. London: Batsford.

Else-Quest, N. M., Hyde, J. S., and Linn, M. C. (2010). Cross-national patterns of gender differences in mathematics: a meta-analysis. *Psychological Bulletin, 136* (1), 103–127.

Ericsson, K. A. (2006). The influence of experience and deliberate practice on the development of superior expert performance. In K. A. Ericsson, N. Charness, P. Feltovich, and R. R. Hoffman (eds.), *The Cambridge Handbook of Expertise and Expert Performance*. Cambridge: Cambridge University Press, 685–706.

Ericsson, K. A., and Charness, N. (1994). Expert performance: its structure and acquisition. *American Psychologist, 49*(8), 725–747.

Ericsson, K. A., Charness, N., Feltovich, P. J., and Hoffman, R. R. (eds.) (2006). *The Cambridge Handbook of Expertise and Expert Performance*. Cambridge: Cambridge University Press.

Ericsson, K. A., and Kintsch, W. (1995). Long-term working memory. *Psychological Review, 102*(2), 211–245.

Ericsson, K. A., Krampe, R. T., and Tesch-Römer, C. (1993). The role of deliberate practice in the acquisition of expert performance. *Psychological Review, 100*(3), 363–406.

Ericsson, K. A., and Lehman, A. C. (1996). Expert and exceptional performance: evidence of maximal adaptation to task constraints. *Annual Review of Psychology, 47*, 273–305.

Ericsson, K. A., and Moxley, J. H. (2012). A critique of Howard's argument for innate limits in chess performance or why we need an account based on acquired skill and deliberate practice. *Applied Cognitive Psychology, 26*(4), 649–653.

Espejo, J., Day, E. A., and Scott, G. (2005). Performance evaluations, need for cognition, and the acquisition of a complex skill. *Personality and Individual Differences, 38*(8), 1867–1877.

Eysenck, H. J. (1952). *The Scientific Study of Personality*. London: Routledge & Kegan Paul.

Eysenck, H. J. (1994). Personality: biological foundations. In P. A. Vernon (ed.), *The Neuropsychology of Individual Differences*. San Diego: Academic Press, 151–207.

Eysenck, H. J., and Eysenck, M. W. (1985). *Personality and Individual Differences: A Natural Science Approach*. New York: Plenum.

Eysenck, H. J., and Eysenck, S. B. G. (1984). *EPQ: Cuestionario de personalidad para niños y adultos*. Barcelona: TEA Ediciones.

Fagan, J. F., and Detterman, D. K. (1992). The Fagan Test of Infant Intelligence: a technical summary. *Journal of Applied Developmental Psychology, 13*(2), 173–193.

Fair, R. C. (2007). Estimated age effects in athletic events and chess. *Experimental Aging Research, 33*(1), 37–57.

Feingold, A. (1994a). Gender differences in personality: a meta-analysis. *Psychological Bulletin, 116*(3), 429–456.

Feingold, A. (1994b). Gender differences in variability in intellectual abilities: a cross-cultural perspective. *Sex Roles, 30*(1), 81–92.

Feingold, A. (1996). Cognitive gender differences: where are they, and why are they there? *Learning and Individual Differences, 8*(1), 25–32.

Fenner, T., Levene, M., and Loizou, G. (2012). A discrete evolutionary model for chess players' ratings. *IEEE Transactions on Computational Intelligence and AI in Games, 4*(2), 84–93.

Ferguson, G. A. (1954). On learning and human ability. *Canadian Journal of Psychology, 8*(2), 95–112.

Ferguson, G. A. (1956). On transfer and the abilities of man. *Canadian Journal of Psychology, 10*(3), 121–131.

Fernández-Amigo, J. (2008). Utilización de material didáctico con recursos de ajedrez para la enseñanza de las matemáticas: estudio de sus efectos sobre una muestra de alumnos de segundo de primaria. Unpublished doctoral dissertation, Universitat Autònoma de Barcelona.

Fernandez-Slezak, D., and Sigman, M. (2012). Do not fear your opponent: sub-optimal changes of a prevention strategy when facing stronger opponents. *Journal of Experimental Psychology: General, 141*(3), 527–538.

Ferrari, V., Didierjean, A., and Marmèche, E. (2008). Effect of expertise acquisition on strategic perception: the example of chess. *Quarterly Journal of Experimental Psychology, 61*(8), 1265–1280.

Ferreira, D., and Palhares, P. (2008). Chess and problem solving involving patterns. *The Mathematics Enthusiast, 5*(2), 249–256.

Fink, A., and Neubauer, A. C. (2006). EEG alpha oscillations during the performance of verbal creativity tasks: differential effects of sex and verbal intelligence. *International Journal of Psychophysiology, 62*(1), 46–53.

Flynn, J. R. (1984). The mean IQ of Americans: massive gains 1932 to 1978. *Psychological Bulletin, 95*(1), 29–51.

Flynn, J. R. (1987). Massive IQ gains in 14 nations: what IQ tests really measure. *Psychological Bulletin, 101*(2), 171–191.

Fond, G., Micoulaud-Franchi, J. A., Brunel, L., Macgregor, A., Miot, S., Lopez, R., et al. (2015). Innovative mechanisms of action for pharmaceutical cognitive enhancement: a systematic review. *Psychiatry Research, 229*(1/2), 12–20.

Forrest, D., Davidson, D., Shucksmith, J., and Glendinning, T. (2005). *Chess Development in Aberdeen's Primary Schools: A Study of Literacy and Social Capital.* Aberdeen: Aberdeen City Council, University of Aberdeen, and Scottish Executive Education Department.

Fox, K., Ford, I., Steg, P. G., Tendera, M., Robertson, M., and Ferrari, R. (2008). Heart rate as a prognostic risk factor in patients with coronary artery disease and left-ventricular systolic dysfunction (BEAUTIFUL): a subgroup analysis of a randomised controlled trial. *The Lancet, 372*, 817–821.

Fox, M. D., Snyder, A. Z., Vincent, J. L., Corbetta, M., Van Essen, D. C., and Raichle, M. E. (2005). The human brain is intrinsically organized into dynamic, anticorrelated functional networks. *Proceedings of the National Academy of Sciences, 102*(27), 9673–9678.

Franic, S., Dolan, C. V., Borsboom, D., Hudziak, J. J., van Beijsterveldt, C. E. M., and Boomsma, D. I. (2013). Can genetics help psychometrics? Improving dimensionality assessment through genetic factor modeling. *Psychological Methods, 18*(3), 406–433.

Frank, A., and D'Hondt, W. (1979). Aptitudes et apprentissage du jeu d'échecs du Zaïre. *Psychopathologie Africaine Dakar, 15*(1), 81–98.

Franke, A. G., Gränsmarck, P., Agricola, A., Schühle, K., Rommel, T., Sebastian, A., et al. (2017). Methylphenidate, modafinil, and caffeine for cognitive enhancement in chess: a double-blind, randomised controlled trial. *European Neuropsychopharmacology, 27*(3), 248–260.

Freund, A. M., and Keil, A. (2013). Out of mind, out of heart: attention affects duration of emotional experience. *Cognition and Emotion, 27*(3), 549–557.

Frey, P. W., and Adesman, P. (1976). Recall memory for visually presented chess positions. *Memory and Cognition, 4*(5), 541–547.

Frydman, M., and Lynn, R. (1992). The general intelligence and spatial abilities of gifted young Belgian chess players. *British Journal of Psychology, 83*(2), 233–235.

Fuentes, J. P., Villafaina, S., Collado-Mateo, D., de la Vega, R., Gusi, N., and Clemente-Suárez, V. J. (2018). Use of biotechnological devices in the quantification of psychophysiological workload of professional chess players. *Journal of Medical Systems, 42*(3), article 40.

Furley, P., and Wood, G. (2016). Working memory, attentional control, and expertise in sports: a review of current literature and directions for future research. *Journal of Applied Research in Memory and Cognition, 5*(4), 415–425.

Furnham, A. (1996). The big five versus the big four: the relationship between the Myers–Briggs Type Indicator (MBTI) and NEO-PI five factor model of personality. *Personality and Individual Differences, 21*(2), 303–307.

Galton, F. (1869). *Hereditary Genius: An Inquiry into Its Laws and Consequences.* London: Macmillan.

Garach, Z., Zaytseva, Y., Kapranova, A., Fiala, O., Horacek, J., Shmukler, A., et al. (2015). EEG correlates of a mental arithmetic task in patients with first episode schizophrenia and schizoaffective disorder. *Clinical Neurophysiology, 126*(11), 2090–2098.

García, H., and Blanch, A. (2016). Tecnochess: una propuesta didáctica para trabajar las competencias lingüística, tecnológica, y matemática mediate el juego del ajedrez y las TIC. *Revista Interuniversitaria de Investigación en Tecnología Educativa, 1*, 39–51.

Garcia, N. V. (2008). Scholastic chess club participation and the academic achievement of Hispanic fifth grade students in south Texas. Unpublished doctoral dissertation, University of Houston.

Gardner, H. (1993). *Frames of Mind: The Theory of Multiple Intelligences.* New York: Basic Books.

Gaschler, R., Progscha, J., Smallbone, K., Ram, N., and Bilalić, M. (2014). Playing off the curve: testing quantitative predictions of skill acquisition theories in development of chess performance. *Frontiers in Psychology, 5*, article 923.

Geary, D. C. (1995a). Reflections of evolution and culture in children's cognition: implications for mathematical development and instruction. *American Psychologist, 50*(1), 24–37.

Geary, D. C. (1995b). Sexual selection and sex differences in spatial cognition. *Learning and Individual Differences, 7*(4), 289–301.

Geary, D. C. (1996). Sexual selection and sex differences in mathematical abilities. *Behavioral and Brain Sciences, 19*(2), 229–284.

Geary, D. C. (1999). Sex differences in mathematical abilities: commentary on the math-fact retrieval hypothesis. *Contemporary Educational Psychology, 24*(3), 267–274.

Geary, D. C. (2006). Sexual selection and the evolution of human sex differences. *Psychological Topics, 15*(2), 203–238.

Geary, D. C. (2010). *Male, Female: The Evolution of Human Sex Differences*, 2nd edn. Washington, DC: American Psychological Association.

Gerdes, C., and Gränsmark, P. (2010). Strategic behavior across gender: a comparison of female and male expert chess players. *Labour Economics, 17* (5), 766–775.

Glickman ME (1999). "Parameter estimation in large dynamic paired comparison experiments," Applied Statistics, 48, 377–394.

Glickman ME (2001). "Dynamic paired comparison models with stochastic variances." Journal of Applied Statistics, 28, 673–689.

Glickman, M. E. (1995). A comprehensive guide to chess ratings. *American Chess Journal, 3*, 59–102.

Glickman, M. E., and Chabris, C. F. (1996). Using chess ratings as data in psychological research. Unpublished manuscript, Boston University, Department of Mathematics and Statistics.

Glickman, M. E., and Jones, A. C. (1999). Rating the chess rating system. *Chance, 12*(2), 21–28.

Gliga, F., and Flesner, P. I. (2014). Cognitive benefits of chess training in novice children. *Procedia – Social and Behavioral Sciences, 116*, 962–967.

Gobet, F. (1992). Learned helplessness in chess players: the importance of task similarity and the role of skill. *Psychological Research, 54*(1), 38–43.

Gobet, F. (1997). A pattern-recognition theory of search in expert problem solving. *Thinking and Reasoning, 3*(4), 291–313.

Gobet, F. (1998). Expert memory: a comparison of four theories. *Cognition, 66*(2), 115–152.

Gobet, F. (2016). *Understanding Expertise: A Multi-Disciplinary Approach*. London: Red Globe Press.

Gobet, F. (2019). *The Psychology of Chess*. Abingdon: Routledge.

Gobet, F., and Campitelli, G. (2006). Educational benefits of chess instruction: a critical review. In T. Redman (ed.), *Chess and Education: Selected Essays from the Koltanowski Conference*. Dallas: Chess Program at the University of Texas at Dallas, 124–143.

Gobet, F., and Campitelli, G. (2007). The role of domain-specific practice, handedness, and starting age in chess. *Developmental Psychology*, *43*(1), 159–172.

Gobet, F., Campitelli, G., and Waters, A. J. (2002). Rise of human intelligence: comments on Howard (1999). *Intelligence*, *20*(4), 303–311.

Gobet, F., and Charness, N. (2006). Expertise in chess. In K. A. Ericsson, N. Charness, P. J. Feltovich, and R. R. Hoffman (eds.), *The Cambridge Handbook of Expertise and Expert Performance*. New York: Cambridge University Press, 523–538.

Gobet, F., and Chassy, P. (2008). Season of birth and chess expertise. *Journal of Biosocial Science*, *40*(2), 313–316.

Gobet, F., and Clarkson, G. (2004). Chunks in expert memory: evidence for the magical number four . . . or is it two? *Memory*, *12*(6), 732–747.

Gobet, F., and Ereku, M. H. (2014). Checkmate to deliberate practice: the case of Magnus Carlsen. *Frontiers in Psychology*, *5*, article 878.

Gobet, F., and Simon, H. A. (1996a). Recall of random and distorted chess positions: implications for the theory of expertise. *Memory and Cognition*, *24* (4), 493–503.

Gobet, F., and Simon, H. A. (1996b). The roles of recognition processes and look-ahead search in time-constrained expert problem solving: evidence from grand-master-level chess. *Psychological Science*, *7*(1), 52–55.

Gobet, F., and Simon, H. A. (1996c). Templates in chess memory: a mechanism for recalling several boards. *Cognitive Psychology*, *31*(1), 1–40.

Gobet, F., and Simon, H. A. (1998a). Expert chess memory: revisiting the chunking hypothesis. *Memory*, *6*(3), 225–255.

Gobet, F., and Simon, H. A. (1998b). Pattern recognition makes search possible: comments on Holding (1992). *Psychological Research*, *61*(3), 204–208.

Gobet, F., and Simon, H. A. (2000). Five seconds or sixty? Presentation time in expert memory. *Cognitive Science*, *24*(4), 651–682.

Gobet, F., and Waters, A. J. (2003). The role of constraints in expert memory. *Journal of Experimental Psychology: Learning, Memory, and Cognition*, *29*(6), 1082–1094.

Goldberg, L. R. (1990). An alternative 'description of personality': the big five factor structure. *Journal of Personality and Social Psychology*, *59*(6), 1216–1229.

Goldin, S. (1978). Effects of orienting tasks on recognition of chess positions. *American Journal of Psychology*, *91*(4), 659–671.

Goldin, S. (1979). Recognition memory for chess positions: some preliminary research. *American Journal of Psychology*, *92*(1), 19–31.

Golf, S. (2015a). Biochemistry and psychology of chess and classical physical exercise: concurring or conflicting evidence? *Journal of Sports Medicine and Doping Studies*, *5*(2), article 158.

Golf, S. (2015b). Doping for chess performance. *Journal of Sports Medicine and Doping Studies*, *5*(3), article 160.

Gondra, J. M. (1994). Juan Huarte de San Juan y las diferencias de inteligencia. *Anuario de Psicología*, *60*, 13–34.

Gong, Y., Ericsson, K. A., and Moxley, J. H. (2015). Recall of briefly presented chess positions and its relation to chess skill. *PLOS ONE*, *10*(3), e0118756.

Gottfredson, L. S. (1997a). Mainstream science on intelligence: an editorial with 52 signatories, history, and bibliography. *Intelligence*, *24*(1), 13–23.

Gottfredson, L. S. (1997b). Why g matters: the complexity of everyday life. *Intelligence*, *24*(1), 79–132.

Gottfredson, L. S. (2004). Intelligence: is it the epidemiologists' elusive 'fundamental cause' of social class inequalities in health? *Journal of Personality and Social Psychology*, *86*(1), 174–199.

Gottschling, J., Spengler, M., Spinath, B., and Spinath, F. M. (2012). The prediction of school achievement from a behavior genetic perspective: results from the German twin study on cognitive ability, self-reported motivation, and school achievement (CoSMoS). *Personality and Individual Differences*, *53*(4), 381–386.

Graber, R. S. (2009). Business lessons from chess: a discussion of parallels between chess strategy and business strategy, and how chess can have applications for business education. *Academy of Educational Leadership Journal*, *13*(1), 79–85.

Grabner, R. H. (2014a). The role of intelligence for performance in the prototypical expertise domain of chess. *Intelligence*, *45*, 26–33.

Grabner, R. H. (2014b). Going beyond the expert-performance framework in the domain of chess. *Intelligence*, *45*, 109–111.

Grabner, R. H., Fink, A., Stipacek, A., Neuper, C., and Neubauer, A. C. (2004). Intelligence and working memory systems: evidence of neural efficiency in alpha band ERD. *Cognitive Brain Research*, *20*(2), 212–225.

Grabner, R. H., Neubauer, A. C., and Stern, E. (2006). Superior performance and neural efficiency: the impact of intelligence and expertise. *Brain Research Bulletin*, *69*(4), 422–439.

Grabner, R. H., Stern, E., and Neubauer, A. C. (2007). Individual differences in chess expertise: a psychometric investigation. *Acta Psychologica*, *124*(3), 398–420.

Gränsmark, P. (2012). Masters of our time: impatience and self-control in high-level chess games. *Journal of Economic Behavior and Organization*, *82*(1), 179–191.

Grau-Pérez, G., and Moreira, K. (2017). A study of the influence of chess on the executive functions in school-aged children. *Estudios de Psicología*, *38*(1), 1–22.

Gray, J. A. (1987). *The Psychology of Fear and Stress*, 2nd edn. Cambridge: Cambridge University Press.

Gross, J. J., and John, O. P. (2003). Individual differences in two emotion regulation processes: implications for affect, relationships, and well-being. *Journal of Personality and Social Psychology*, *85*(2), 348–362.

Grudnik, J. L., and Kranzler, J. H. (2001). Meta-analysis of the relationship between intelligence and inspection time. *Intelligence*, *29*(6), 523–535.

Guida, A., Gobet, F., Tardieu, H., and Nicolas, S. (2012). How chunks, long-term working memory and templates offer a cognitive explanation for neuroimaging data on expertise acquisition: a two-stage framework. *Brain and Cognition*, *79* (3), 221–244.

Guilford, J. P. (1975). Factors and factors of personality. *Psychological Bulletin*, *82* (5), 802–814.

Günther, F., Wawro, N., and Bammann, K. (2009). Neural networks for modeling gene–gene interactions in association studies. *BMC Genetics*, *10*, article 87.

Gustafsson, J. E. (1978). A note on class effects in aptitude X treatment interactions. *Journal of Educational Psychology*, *70*(2), 142–146.

Haier, R. J., Jung, R. E., Yeo, R. A., Head, K., and Alkire, M. T. (2005). The neuroanatomy of general intelligence: sex matters. *NeuroImage*, *25*(1), 320–327.

Hair, J. F., Black, W. C., Babin, B. J., and Anderson, R. E. (2010). *Multivariate Data Analysis*. Englewood Cliffs, NJ: Prentice Hall.

Halpern, D. F. (1996). Changing data, changing minds: what the data on cognitive sex differences tell us and what we hear. *Learning and Individual Differences*, *8* (1), 73–82.

Halpern, D. F. (1997). Sex differences in intelligence: implications for education. *American Psychologist*, *52*(10), 1091–1102.

Halpern, D. F., Benbow, C. P., Geary, D. C., Gur, R. C., Hyde, J. S., and Gernsbacher, M. A. (2007). The science of sex differences in science and mathematics. *Psychological Science in the Public Interest*, *8*(1), 1–51.

Halpern, D. F., and Wright, T. M. (1996). A process-oriented model of cognitive sex differences. *Learning and Individual Differences*, *8*(1), 3–24.

Hambrick, D. Z., Meinz, E. J., and Oswald, F. L. (2007). Individual differences in current events knowledge: contributions of ability, personality, and interests. *Memory and Cognition*, *35*(2), 304–316.

Hambrick, D. Z., Oswald, F. L., Altmann, E. M., Meinz, E. J., Gobet, F., and Campitelli, G. (2014). Deliberate practice: is that all it takes to become an expert? *Intelligence*, *45*, 34–45.

Hanggi, J., Brütsch, K., Siegel, A. M., and Jancke, L. (2014). The architecture of the chess player's brain. *Neuropsychologia*, *62*(1), 152–162.

Harding, T. (2009). Battle at long range: correspondence chess in Britain and Ireland, 1824–1914, a social and cultural history. Unpublished PhD, Trinity College Dublin.

Hedges, L. V., and Nowell, A. (1995). Sex differences in mental test scores, variability, and numbers of high-scoring individuals. *Science, 269*, 41–45.

Herbrich, R., and Graepel, T. (2006). TrueSkillTM: a Bayesian skill rating system, Technical Report MSR-TR-2006–80. Redmond, WA: Microsoft.

Hernández, P., and Rodríguez, H. (2006). Success in chess mediated by mental molds. *Psicothema, 18*(4), 704–710.

Herrmann, C. S., Fründ, I., and Lenz, D. (2010). Human gamma-band activity: a review on cognitive and behavioral correlates and network models. *Neuroscience and Biobehavioral Reviews, 34*(7), 981–992.

Herrnstein, R. J., and Murray, C. (1994). *The Bell Curve: Intelligence and Class Structure in American Life*. New York: Free Press.

Hill, N. M., and Schneider, W. (2006). Brain changes in the development of expertise: neuroanatomical and neurophysiological evidence about skill-based adaptations. In K. A. Ericsson, N. Charness, P. J. Feltovich, and R. R. Hoffman (eds.), *The Cambridge Handbook of Expertise and Expert Performance*. New York: Cambridge University Press, 653–682.

Hodgins-Davis, A., and Townsend, J. P. (2009). Evolving gene expression: from G to E to G X E. *Trends in Ecology and Evolution, 24*(12), 649–658.

Holding, D. (1985). *The Psychology of Chess Skill*. Hillsdale, NJ: Lawrence Erlbaum Associates.

Holding, D. (1992). Theories of chess skill. *Psychological Research, 54*(1), 10–16.

Holding, D., and Pfau, H. D. (1985). Thinking ahead in chess. *American Journal of Psychology, 98*(2), 271–282.

Holding, D., and Reynolds, R. (1982). Recall or evaluation of chess positions as determinants of chess skill. *Memory and Cognition, 10*(3), 237–242.

Holland, J. L. (1959). A theory of vocational choice. *Journal of Counseling Psychology, 6*(1), 35–45.

Holland, J. L. (1996). Exploring careers with a typology: what we have learned and some new directions. *American Psychologist, 51*(4), 397–406.

Hong, S., and Bart, W. M. (2007). Cognitive effects of chess instruction on students at risk for academic failure. *International Journal of Special Education, 22*(3), 89–96.

Hopwood, C. J., Donnellan, M. B., Blonigen, D. M., Krueger, R. F., McGue, M., Iacono, W. G., et al. (2011). Genetic and environmental influences on personality trait stability and growth during the transition to adulthood: a three-wave longitudinal study. *Journal of Personality and Social Psychology, 100*(3), 545–556.

Horgan, D. D. (1992). Children and chess expertise: the role of calibration. *Psychological Research, 54*(1), 44–50.

Horgan, D. D., Millis, K., and Neimeyer, R. A. (1989). Cognitive reorganization and the development of chess expertise. *International Journal of Personal Construct Psychology, 2*(1), 15–36.

Horgan, D. D., and Morgan, D. (1990). Chess expertise in children. *Applied Cognitive Psychology, 4*(1), 109–128.

Howard, R. W. (1999). Preliminary real-world evidence that average human intelligence really is rising. *Intelligence*, *27*(3), 235–250.

Howard, R. W. (2005a). Are gender differences in high achievement disappearing? A test in one intellectual domain. *Journal of Biosocial Science*, *37*(3), 371–380.

Howard, R. W. (2005b). Objective evidence of rising population ability: a detailed examination of longitudinal chess data. *Personality and Individual Differences*, *38*(2), 347–363.

Howard, R. W. (2006). A complete database of international chess players and chess performance ratings for varied longitudinal studies. *Behavior Research Methods*, *38*(4), 698–703.

Howard, R. W. (2009). Individual differences in expertise development over decades in a complex intellectual domain. *Memory and Cognition*, *37*(2), 194–209.

Howard, R. W. (2011a). Does high-level intellectual performance depend on practice alone? Debunking the Polgar sisters case. *Cognitive Development*, *26* (3), 196–202.

Howard, R. W. (2011b). Testing the accuracy of the retrospective recall method used in expertise research. *Behavior Research Methods*, *43*(4), 931–941.

Howard, R. W. (2012a). Longitudinal effects of different types of practice on the development of chess expertise. *Applied Cognitive Psychology*, *26*(3), 359–369.

Howard, R. W. (2012b). Selecting the better model by using data, Occam's razor, and a little common sense: reply to Ericsson. *Applied Cognitive Psychology*, *26* (4), 654–656.

Howard, R. W. (2013). Practice other than playing games apparently has only a modest role in the development of chess expertise. *British Journal of Psychology*, *104*(1), 39–56.

Howard, R. W. (2014a). Explaining male predominance at the apex of intellectual achievement. *Personality and Individual Differences*, *68*, 217–220.

Howard, R. W. (2014b). Gender differences in intellectual performance persist at the limits of individual capabilities. *Journal of Biosocial Science*, *46*(3), 386–404.

Howard, R. W. (2014c). Learning curves in highly skilled chess players: a test of the generality of the power law of practice. *Acta Psychologica*, *151*, 16–23.

Howe, M. J. A., Davidson, J. W., and Sloboda, J. A. (1998). Innate talents: reality or myth? *Behavioral and Brain Sciences*, *21*(3), 399–442.

Hu, L., and Bentler, P. M. (1999). Cutoff criteria for fit indexes in covariance structure analysis: conventional criteria versus new alternatives. *Structural Equation Modeling*, *6*(1), 1–55.

Huarte de San Juan, J. (1593). *Examen de ingenios para las sciencias*. Alicante: Biblioteca Virtual Miguel de Cervantes.

Hunt, E. (2011). *Human Intelligence*. New York: Cambridge University Press.

Hunt, E., and Carlson, J. (2007). Considerations relating to the study of group differences in intelligence. *Perspectives on Psychological Science, 2*(2), 194–213.

Hyde, J. S. (2005). The gender similarities hypothesis. *American Psychologist, 60*(6), 581–592.

Ilies, R., Arvey, R., and Bouchard, T. J. (2006). Darwinism, behavioral genetics, and organizational behavior: a review and agenda for future research. *Journal of Organizational Behavior, 27*(2), 121–141.

Ilies, R., and Judge, T. A. (2003). On the heritability of job satisfaction: the mediating role of personality. *Journal of Applied Psychology, 88*(4), 750–759.

Irwing, P., and Lynn, L. (2005). Sex differences in means and variability on the progressive matrices in university students: a meta-analysis. *British Journal of Psychology, 96*(4), 505–524.

Jastrzembski, T. S., Charness, N., and Vasyukova, C. (2006). Expertise and age effects on knowledge activation in chess. *Psychology and Aging, 21*(2), 401–405.

Jensen, A. R., and Munro, E. (1979). Reaction time, movement time, and intelligence. *Intelligence, 3*(2), 121–126.

Jerrim, J., Macmillan, L., Micklewright, J., Sawtell, M., and Wiggins, M. (2016). *Chess in Schools: Evaluation Report and Executive Summary*. London: Education Endowment Foundation.

Johnson, W., Bouchard, T. J., McGue, M., Segal, N. L., Tellegen, A., Keyes, M., et al. (2007). Genetic and environmental influences on the verbal–perceptual–image rotation (VPR) model of the structure of mental abilities in the Minnesota study of twins reared apart. *Intelligence, 35*(6), 542–562.

Johnson, W., Carothers, A., and Deary, I. J. (2008). Sex differences in variability in general intelligence: a new look at the old question. *Perspectives on Psychological Science, 3*(6), 518–531.

Johnson, W., and Deary, I. J. (2011). Placing inspection time, reaction time, and perceptual speed in the broader context of cognitive ability: the VPR model in the Lothian Birth Cohort 1936. *Intelligence, 39*(5), 405–417.

Johnson, W., Deary, I. J., and Iacono, W. G. (2009). Genetic and environmental transactions underlying educational attainment. *Intelligence, 37*(5), 466–478.

Joireman, J. A., Fick, C. S., and Anderson, J. W. (2002). Sensation seeking and involvement in chess. *Personality and Individual Differences, 32*(3), 509–515.

Jorgensen, R. S., Johnson, B. T., Kolodziej, M. E., and Schreer, G. E. (1996). Elevated blood pressure and personality: a meta-analytic review. *Psychological Bulletin, 120*(2), 293–320.

Joseph, E., Easvaradoss, V., and Solomon, N. J. (2016). Impact of chess training on academic performance of rural Indian school children. *Open Journal of Social Sciences, 4*(2), 20–24.

Juan-Espinosa, M., and García Rodríguez, L. (2004). *Nuestra personalidad: En qué y por qué somos diferentes*. Madrid: Biblioteca Nueva.

Judge, T. A., Heller, D., and Mount, M. K. (2002). Five-factor model of personality and job satisfaction: a meta-analysis. *Journal of Applied Psychology, 87*(3), 530–541.

Jung, R. E., and Haier, R. J. (2007). The parieto-frontal integration theory (P-FIT) of intelligence: converging neuroimaging evidence. *Behavioral and Brain Sciences, 30*(2), 135–187.

Kahneman, D. (2011). *Thinking, Fast and Slow*. London: Penguin Books.

Kanai, R., and Rees, G. (2011). The structural basis of inter-individual differences in human behaviour and cognition. *Nature Reviews: Neuroscience, 12*(4), 231–242.

Kaufman, A. S., and Kaufman, N. L. (1993). *Manual: Kaufman Adolescent and Adult Intelligence Test*. Circle Pines, MN: American Guidance Service.

Kazemi, F., Yektayar, M., and Abad, A. M. B. (2012). Investigation: the impact of chess play on developing meta-cognitive ability and math problem-solving power of students at different levels of education. *Procedia – Social and Behavioral Sciences, 32*, 372–379.

Kell, H. J., Lubinski, D., and Benbow, C. P. (2013). Who rises to the top? Early indicators. *Psychological Science, 24*(5), 648–659.

Kelly, E. J. (1985). The personality of chessplayers. *Journal of Personality Assessment, 49*(3), 282–284.

Kenemans, L. (2013). *A Primer on EEG and Related Measures of Brain Activity*. Utrecht: Department of Experimental Psychology and Psychopharmacology, University of Utrecht.

Kennis, M., Rademaker, A. R., and Geuze, E. (2013). Neural correlates of personality: an integrative review. *Neuroscience and Biobehavioral Reviews, 37*(1), 73–95.

Kiesel, A., Kunde, W., Pohl, C., Berner, M., and Hoffmann, J. (2009). Playing chess unconsciously. *Journal of Experimental Psychology: Learning, Memory, and Cognition, 35*(1), 292–298.

Kimura, D. (1993). Sex differences in the brain. In *Mind and Brain: Readings from Scientific American Magazine*. New York: W. H. Freeman, 79–89.

Kimura, D. (1996). Sex, sexual orientation and sex hormones influence human cognitive function. *Current Opinion in Neurobiology, 6*(2), 259–263.

Klein, G., and Peio, K. (1989). Use of a prediction paradigm to evaluate proficient decision making. *American Journal of Psychology, 102*(3), 321–331.

Klein, G., Wolf, S., Militello, L., and Zsambok, C. (1995). Characteristics of skilled option generation in chess. *Organizational Behavior and Human Decision Processes, 62*(1), 63–69.

Knapp, M. (2010). Are participation rates sufficient to explain gender differences in chess performance? *Proceedings of the Royal Society B, 277*, 2269–2270.

Kotov, A. (1971). *Think Like a Grandmaster*. London: Batsford.

Kotov, R., Gamez, W., Schmidt, F., and Watson, D. (2010). Linking 'big' personality traits to anxiety, depressive, and substance use disorders: a meta-analysis. *Psychological Bulletin, 136*(5), 768–821.

Krapohla, E., Rimfeld, K., Shakeshaft, N. G., Trzaskowski, M., McMillan, A., Pingault, J. B., et al. (2014). The high heritability of educational achievement reflects many genetically influenced traits, not just intelligence. *PNAS*, *111*(42), 15273–15278.

Krawczyk, D. C., Boggan, A. L., McClelland, M. M., and Bartlett, J. C. (2011). The neural organization of perception in chess experts. *Neuroscience Letters*, *499*(2), 64–69.

Krogius, N. (1976). *Psychology in Chess*. London: RHM Press.

Lane, D., and Robertson, L. (1979). The generality of the levels of processing hypothesis: an application to memory for chess positions. *Memory and Cognition*, *7*(4), 253–256.

Lassiter, G. D. (2000). The relative contributions of recognition and search-evaluation processes to high-level chess performance: comment on Gobet and Simon. *Psychological Science*, *11*(2), 172–173.

Leone, M. J., Petroni, A., Fernández-Slezak, D., and Sigman, M. (2012). The tell-tale heart: heart rate fluctuations index objective and subjective events during a game of chess. *Frontiers in Human Neuroscience*, *6*, article 273.

Levitt, S. D., List, J. A., and Sadoff, S. E. (2011). Checkmate: exploring backward induction among chess players. *American Economic Review*, *101* (2), 975–990.

Li, K., Jiang, J., Qiu, L., Yang, X., Huang, X., Lui, S., et al. (2015). A multimodal MRI dataset of professional chess players. *Scientific Data*, *2*, article 150044.

Lindberg, S. M., Hyde, J. S., Petersen, J. L., and Linn, M. C. (2010). New trends in gender and mathematics performance: a meta-analysis. *Psychological Bulletin*, *136*(6), 1123–1135.

Linhares, A., and Freitas, A. E. T. A. (2010). Questioning Chase and Simon's (1973) 'Perception in chess': the 'experience recognition' hypothesis. *New Ideas in Psychology*, *28*(1), 64–78.

Little, T. D., Schnabel, K. U., and Baumert, J. (1998). *Modeling Longitudinal and Multiple Group Data: Practical Issues, Applied Approaches, and Specific Examples*. Mahwah, NJ: Lawrence Erlbaum Associates.

Llaveria, A., Blanch, A., Aluja, A., and Cornadó, M. P. (2016). Personalidad y motivación en jugadores de ajedrez. Paper presented at the fifth conference of the Asociación Iberoamericana para la Investigación de las Diferencias Individuales, Sitges, Spain, 15 September.

Lohman, D. F., and Hagen, E. P. (2002). *Cognitive Abilities Test (Form 6): Research Handbook*. Itasca, IL: Riverside Publishing.

Lubinski, D. (2000). Scientific and social significance of assessing individual differences: 'sinking shafts at a few critical points'. *Annual Review of Psychology*, *51*, 405–444.

Lubinski, D. (2010). Spatial ability and STEM: a sleeping giant for talent identification and development. *Personality and Individual Differences*, *49*(4), 344–351.

Lubinski, D., and Benbow, C. P. (2006). Study of mathematically precocious youth after 35 years: uncovering antecedents for the development of math-science expertise. *Perspectives on Psychological Science, 1*(4), 316–345.

Lubinski, D., Benbow, C. P., and Kell, H. J. (2014). Life paths and accomplishments of mathematically precocious males and females four decades later. *Psychological Science, 25*(12), 2217–2232.

Lubinski, D., Benbow, C. P., Shea, D. L., Eftekhari-Sanjani, H., and Halvorson, M. B. J. (2001). Men and women at promise for scientific excellence: similarity not dissimilarity. *Psychological Science, 12*(4), 309–317.

Lubinski, D., Webb, R. M., Morelock, M. J., and Benbow, C. P. (2001). Top 1 in 10,000: a 10-year follow-up of the profoundly gifted. *Journal of Applied Psychology, 86*(4), 718–729.

Luque-Casado, A., Zabala, M., Morales, E., Mateo-March, M., and Sanabria, D. (2013). Cognitive performance and heart rate variability: the influence of fitness level. *PLOS ONE, 8*(2), e56935.

Lynn, R., and Irwing, P. (2004). Sex differences on the progressive matrices: a meta-analysis. *Intelligence, 32*(5), 481–498.

Maas, H. L. J., and Spinath, F. M. (2012). Personality and coping with professional demands: a behavioral genetics analysis. *Journal of Occupational Health Psychology, 17*(3), 376–385.

Maass, A., D'Ettole, C., and Cadinu, M. (2008). Checkmate? The role of gender stereotypes in the ultimate intellectual sport. *European Journal of Social Psychology, 38*(2), 231–245.

McCrae, R. R. (2001). Trait psychology and culture: exploring intercultural comparisons. *Journal of Personality and Social Psychology, 69*(6), 819–846.

McCrae, R. R. (2002). The maturation of personality psychology: adult personality development and psychological well-being. *Journal of Research in Personality, 36* (4), 307–317.

McCrae, R. R., and Costa, P. T. (1997). Personality trait structure as a human universal. *American Psychologist, 52*(5), 509–516.

McCrae, R. R., Costa, P. T., Pedroso de Lima, M., Simöes, A., Ostendorf, F., Angleitner, A., et al. (1999). Age differences in personality across the adult life span: parallels in five cultures. *Developmental Psychology, 35*(2), 466–477.

McCrae, R. R., Scally, M., Terracciano, A., Abecasis, G. R., and Costa, P. T. (2010). An alternative to the search for single polymorphisms: toward molecular personality scales for the five-factor model. *Journal of Personality and Social Psychology, 99*(6), 1014–1024.

McGregor, S., and Howes, A. (2002). The role of attack and defense semantics in skilled players' memory for chess positions. *Memory and Cognition, 30*(5), 707–717.

McGrew, K. S. (2009). CHC theory and the human cognitive abilities project: standing on the shoulders of the giants of psychometric intelligence research. *Intelligence, 37*(1), 1–10.

Macnamara, B. N., Hambrick, D. Z., and Oswald, F. L. (2014). Deliberate practice and performance in music, games, sports, education, and professions: a meta-analysis. *Psychological Science, 25*(8), 1608–1618.

Macnamara, B. N., Moreau, D., and Hambrick, D. Z. (2016). The relationship between deliberate practice and performance in sports: a meta-analysis. *Perspectives on Psychological Science, 11*(3), 333–350.

Mardis, E. R. (2009). The impact of next-generation sequencing technology on genetics. *Trends in Genetics, 24*(3), 133–141.

Marmèche, E., and Didierjean, A. (2001). Is generalisation conservative? A study with novices in chess. *European Journal of Cognitive Psychology, 13*(4), 475–491.

Marsella, A. J., Dubanoski, J., Hamada, W. C., and Morse, H. (2000). The measurement of personality across cultures: historical, conceptual, and methodological issues and considerations. *American Behavioral Scientist, 44*(1), 41–62.

Mayer, J. D. (2005). A tale of two visions: can a new view of personality help integrate psychology? *American Psychologist, 60*(4), 294–307.

Mazur, A., Booth, A., and Dabbs, J. M. (1992). Testosterone and chess competition. *Social Psychology Quarterly, 55*(1), 70–77.

Mihailov, E., and Savulescu, J. (2018). Social policy and cognitive enhancement: lessons from chess. *Neuroethics, 11*(2), 115–127.

Mireles, D. E., and Charness, N. (2002). Computational explorations of the influence of structured knowledge on age-related cognitive decline. *Psychology and Aging, 17*(2), 245–259.

Moul, C. C., and Nye, J. V. C. (2009). Did the Soviets collude? A statistical analysis of championship chess 1940–1978. *Journal of Economic Behavior and Organization, 70*(1/2), 10–21.

Moxley, J. H., and Charness, N. (2013). Meta-analysis of age and skill effects on recalling chess positions and selecting the best move. *Psychonomic Bulletin Review, 20*(5), 1017–1022.

Moxley, J. H., Ericsson, A. K., Charness, N., and Krampe, R. T. (2012). The role of intuition and deliberative thinking in experts' superior tactical decision-making. *Cognition, 124*(1), 72–78.

Munafó, M. R., Clark, T. G., Moore, L. R., Payne, E., Walton, R., and Flint, J. (2003). Genetic polymorphisms and personality in healthy adults: a systematic review and meta-analysis. *Molecular Psychiatry, 8*(5), 471–484.

Naglieri, J. A., and Das, J. P. (1997). *Cognitive Assessment System Interpretive Handbook*. Itasca, IL: Riverside Publishing.

Neisser, U., Boodoo, G., Bouchard, T., Boykin, A. W., Brody, N., Ceci, S. J., et al. (1996). Intelligence: knowns and unknowns. *American Psychologist, 51*(2), 77–101.

Nettelbeck, T. (2011). Basic processes of intelligence. In R. J. Sternberg and S. B. Kaufman (eds.), *The Cambridge Handbook of Intelligence*. New York: Cambridge University Press, 371–393.

Nettelbeck, T., and Lally, M. (1976). Inspection time and measured intelligence. *British Journal of Psychology, 67*(1), 17–22.

Neubauer, A. C., and Fink, A. (2009a). Intelligence and neural efficiency. *Neuroscience and Biobehavioral Reviews, 33*(7), 1004–1023.

Neubauer, A. C., and Fink, A. (2009b). Intelligence and neural efficiency: measures of brain activation versus measures of functional connectivity in the brain. *Intelligence, 37*(2), 223–229.

Newell, A., and Rosenbloom, P. S. (1981). Mechanisms of skill acquisition and the law of practice. In J. R. Anderson (ed.), *Cognitive Skills and Their Acquisition*. Hillsdale, NJ: Lawrence Erlbaum Associates, 1–55.

Nichelli, P., Grafman, J., Pietrini, P., Alway, D., Carton, J. D., and Miletich, R. (1994). Brain activity in chess playing. *Nature, 369*, 191.

Nippold, M. A. (2009). School-age children talk about chess: does knowledge drive syntactic complexity? *Journal of Speech, Language, and Hearing Research, 52*(4), 856–871.

Nokes, T. J. (2009). Mechanisms of knowledge transfer. *Thinking and Reasoning, 15*(1), 1–36.

Nosek, B. A., Smyth, F. L., Sriram, N., Lindner, N. M., Devos, T., Ayala, A., et al. (2009). National differences in gender–science stereotypes predict national sex differences in science and math achievement. *Proceedings of the National Academies of Science, 106*(26), 10593–10597.

Onofrj, M., Curatola, L., Valentini, G., Antonelli, M., Thomas, A., and Fulgente, T. (1995). Non-dominant dorsal-prefrontal activation during chess problem solution evidenced by single photon emission computerized tomography (SPECT). *Neuroscience Letters, 198*(3), 169–172.

Palacios-Huerta, I., and Volij, O. (2009). Field centipedes. *American Journal Economic Review, 99*(4), 1619–1635.

Patel, S. H., and Azzam, P. N. (2005). Characterization of N200 and P300: selected studies of the event-related potential. *International Journal of Medical Science, 2* (4), 147–154.

Peralta, F., and de Dovitiis, A. (2007). *Las dos caras del entrenamiento*. Andorra la Vella: Esfera Editorial.

Pervin, L. A. (ed.) (1990). *Handbook of Personality: Theory and Research*. New York: Guilford Press.

Pfau, H. D., and Murphy, M. D. (1988). Role of verbal knowledge in chess skill. *American Journal of Psychology, 101*(1), 73–86.

Pfleger, H., Stocker, K., Pabst, H., and Haralambie, G. (1980). [Sports medical examination of top class chess players]. *Munchener Medizinische Wochenschrift, 122*(28), 1041–1044.

Pinker, S. (2002). *The Blank Slate: The Modern Denial of Human Nature*. London: Penguin Books.

Plomin, R., and Daniels, D. (1987). Why are children in the same family so different from one another? *Behavioral and Brain Sciences, 10*(1), 1–60.

Plomin, R., DeFries, J. C., and Loehlin, J. C. (1977). Genotype–environment interaction and correlation in the analysis of human behavior. *Psychological Bulletin*, *84*(2), 309–322.

Plomin, R., DeFries, J. C., McClearn, G. E., and Rutter, M. (1997). *Behavioral Genetics*, 3rd edn. New York: Freeman.

Plomin, R., Owen, M. J., and McGuffin, P. (1994). The genetic basis of complex human behaviors. *Science*, *264*, 1733–1739.

Plomin, R., and Petrill, S. A. (1997). Genetics and intelligence: what's new? *Intelligence*, *24*(1), 53–77.

Plomin, R., and Rende, R. (1991). Human behavioral genetics. *Annual Review of Psychology*, *42*, 161–190.

Plomin, R., Shakeshaft, N. G., McMillan, A., and Trzaskowski, M. (2014). Nature, nurture, and expertise. *Intelligence*, *45*, 46–59.

Plomin, R., and Spinath, F. M. (2004). Intelligence: genetics, genes, and genomics. *Journal of Personality and Social Psychology*, *86*(1), 112–129.

Poortinga, Y. H., and van Hemert, D. A. (2001). Personality and culture: demarcating between the common and the unique. *Journal of Personality*, *69*(6), 1034–1060.

Poropat, A. (2009). A meta-analysis of the five-factor model of personality and academic performance. *Psychological Bulletin*, *135*(2), 322–338.

Posthuma, D., de Geus, E. J. C., and Boomsma, D. I. (2001). Perceptual speed and IQ are associated through common genetic factors. *Behavior Genetics*, *31*(6), 593–602.

Powell, J. L., Grossi, D., Corcoran, R., Gobet, F., and Garcia-Fiñana, M. (2017). The neural correlates of theory of mind and their role during empathy and the game of chess: a functional magnetic resonance imaging study. *Neuroscience*, *355*, 149–160.

Prediger, D. J. (1982). Dimensions underlying Holland's hexagon: missing link between interests and occupations? *Journal of Vocational Behavior*, *21*(3), 259–287.

Prinzie, P., Stams, G. J. J., Dekovic, M., Reijntjes, A. H. A., and Belsky, J. (2009). The relations between parents' big five personality factors and parenting: a meta-analytic review. *Journal of Personality and Social Psychology*, *97*(2), 351–362.

R Development Core Team. (2015). R: a language and environment for statistical computing. R Foundation for Statistical Computing, 10 February, www.gbif.org/tool/81287/r-a-language-and-environment-for-statistical-computing.

Rakic, P. (2009). Evolution of the neocortex: perspective from developmental biology. *Nature Review Neuroscience*, *10*(10), 724–735.

Ramos, L., Arán, V., and Krumm, G. (2018). Funciones ejecutivas y práctica de ajedrez: un estudio en niños escolarizados. *Psicogente*, *21*(39), 25–34.

Rasbash, J., Jenkins, J., O'Connor, T. G., Tackett, J., and Reiss, D. (2011). A social relations model of observed family negativity and positivity using a genetically informative sample. *Journal of Personality and Social Psychology, 100*(3), 474–491.

Raven, J., & Raven, J. (eds.) (2008). *Uses and Abuses of Intelligence: Studies Advancing Spearman and Raven's Quest for Non-Arbitrary Metrics.* Unionville, NY: Royal Fireworks Press.

Read, S. J., and Miller, L. C. (2002). Virtual personalities: a neural network model of personality. *Personality and Social Psychology Review, 6*(4), 357–369.

Read, S. J., Monroe, B. M., Brownstein, A. L., Yang, Y., Chopra, G., and Miller, L. C. (2010). A neural network model of the structure and dynamics of human personality. *Psychological Review, 117*(1), 61–92.

Reilly, D., Neumann, D. L., and Andrews, G. (2015). Sex differences in mathematics and science achievement: a meta-analysis of National Assessment of Educational Progress assessments. *Journal of Educational Psychology, 107*(3), 645–662.

Reingold, M. E., Charness, N., Pomplun, M., and Stampe, M. D. (2001). Visual span in expert chess players: evidence from eye movements. *Psychological Science, 12*(1), 48–55.

Rennig, J., Bilalić, M., Huberle, E., Kamath, H. O., and Himmelbach, M. (2013). The temporo-parietal junction contributes to global gestalt-perception: evidence from studies in chess experts. *Frontiers in Human Neuroscience, 7*, article 513.

Repantis, D., Schlattmann, P., Laisney, O., and Heuser, I. (2010). Modafinil and methylphenidate for neuroenhancement in healthy individuals: a systematic review. *Pharmacological Research, 62*(3), 187–206.

Revelle, W., Wilt, J., and Condon, D. M. (2011). Individual differences and differential psychology: a brief history and prospect. In T. Chamorro-Premuzic, S. von Stumm, and A. Furnham (eds.), *The Wiley-Blackwell Handbook of Individual Differences.* Chichester: Blackwell Publishing, 3–38.

Reynolds, R. (1982). Search heuristics of chess players of different calibers. *American Journal of Psychology, 95*(3), 383–392.

Reynolds, R. (1992). Recognition of expertise in chess players. *American Journal of Psychology, 105*(3), 409–415.

Rieder, M. K., Rahm, B., Williams, J. D., and Kaiser, J. (2011). Human gamma-band activity and behavior. *International Journal of Psychophysiology, 79*(1), 39–48.

Rietveld, C. A., Medland, S. E., Derringer, J., Yang, J., Esko, T., Martin, N. W., et al. (2013). GWAS of 126,559 individuals identifies genetic variants associated with educational attainment. *Science, 340*, 1467–1471.

Rifner, P. J. (1992). Playing chess: a study of the transfer of problem-solving skills in students with average and above average intelligence. Unpublished doctoral dissertation, Purdue University, West Lafayette, IN.

Robbins, T., Anderson, J., Barker, D., Bradley, C., Fearnyhough, C., Henson, R., et al. (1996). Working memory in chess. *Memory and Cognition, 24*(1), 83–93.

Roberts, B. W., Caspi, A., and Moffit, T. E. (2003). Work experiences and personality development in young adulthood. *Journal of Personality and Social Psychology*, *84*(3), 582–593.

Roberts, B. W., and DelVecchio, W. F. (2000). The rank-order consistency of personality traits from childhood to old age: a quantitative review of longitudinal studies. *Psychological Bulletin*, *126*(1), 3–25.

Roberts, B. W., Kuncel, N. R., Shiner, R., Caspi, A., and Goldberg, L. R. (2012). The power of personality: the comparative validity of personality traits, socioeconomic status, and cognitive ability for predicting important life outcomes. *Perspectives on Psychological Science*, *2*(4), 313–345.

Roid, G. H. (2003). *Stanford–Binet Intelligence Scales*, 5th edn. Itasca, IL: Riverside Publishing.

Rojas Vidaurreta, L. (2011). Aproximación al estudio de la flexibilidad cognitiva en niños ajedrecistas. *Revista Cubana de Medicina del Deporte y la Cultura Física*, *6*, article 1728-922X.

Roring, R. W., and Charness, N. (2007). A multilevel model analysis of expertise in chess across the life span. *Psychology and Aging*, *22*(2), 291–299.

Rosenberg, S., Nelson, C., and Vivekananthan, P. S. (1968). A multidimensional approach to the structure of personality impressions. *Journal of Personality and Social Psychology*, *9*(4), 283–294.

Rosenthal, R. (1995). Writing meta-analytic reviews. *Psychological Bulletin*, *118*(2), 183–192.

Rosholm, M., Mikkelsen, M. B., and Gumede, K. (2017). Your move: the effect of chess on mathematic test scores. *PLOS ONE*, *12*(5), e0177257.

Rosseel, Y. (2012). lavaan: an R package for structural equation modeling. *Journal of Statistical Software*, *48*(2), 1–36.

Rothgerber, H., and Wolsiefer, K. (2014). A naturalistic study of stereotype threat in young female chess players. *Group Processes and Intergroup Relations*, *17*(1), 79–90.

Rubin, E. (1960). The age factor in master chess. *The American Statistician*, *14*(5), 19–21.

Ruigrok, A. N. V., Salimi-Khorshidi, G., Lai, M. C., Baron-Cohen, S., Lombardo, M. V., Tait, R. J., et al. (2014). A meta-analysis of sex differences in human brain science. *Neuroscience and Biobehavioral Reviews*, *39*(1), 34–50.

Ruiz, F., and Luciano, C. (2009). Eficacia de la terapia de aceptación y compromiso (ACT) en la mejora del rendimiento ajedrecístico de jóvenes promesas. *Psicothema*, *21*(3), 347–352.

Rushton, J. P., Bons, T. A., Vernon, P. A., and Čvorović, J. (2007). Genetic and environmental contributions to population group differences on the Raven's progressive matrices estimated from twins reared together and apart. *Proceedings of the Royal Society B*, *274*, 1773–1777.

Saariluoma, P. (1985). Chess players' intake of task-relevant cues. *Memory and Cognition*, *13*(5), 385–391.

Saariluoma, P. (1995). *Chess Players' Thinking: A Cognitive Psychological Approach*. London: Routledge.

Saariluoma, P. (2001). Chess and content-oriented psychology of thinking. *Psicológica*, *22*(1), 143–164.

Saariluoma, P., and Kalakoski, V. (1997). Skilled imagery and long-term working memory. *American Journal of Psychology*, *110*(2), 177–201.

Saariluoma, P., Karlsson, H., Lyytinen, H., Teräs, M., and Geisler, F. (2004). Visuospatial representations used by chess experts: a preliminary study. *European Journal of Cognitive Psychology*, *16*(5), 753–766.

Sala, G., Burgoyne, A. P., Macnamara, B. N., Hambrick, D. Z., Campitelli, G., and Gobet, F. (2017). Checking the 'academic selection' argument. Chess players outperform non-chess players in cognitive skills related to intelligence: a meta-analysis. *Intelligence*, *61*, 130–139.

Sala, G., Foley, J. P., and Gobet, F. (2017). The effects of chess instruction on pupils' cognitive and academic skills: state of the art and theoretical challenges. *Frontiers in Psychology*, *8*, article 238.

Sala, G., and Gobet, F. (2016). Do the benefits of chess instruction transfer to academic and cognitive skills? A meta-analysis. *Educational Research Review*, *18*, 16–57.

Sala, G., and Gobet, F. (2017). Does chess instruction improve mathematical problem-solving ability? Two experimental studies with an active control group. *Learning and Behavior*, *45*(4), 414–421.

Sala, G., Gorini, A., and Pravettoni, G. (2015). Mathematical problem-solving abilities and chess: an experimental study on young pupils. *Sage Open*, *5*, 1–9.

Samarian, S. (2008). *Das systematische Schachtraining: Trainingsmethoden, Strategien und Kombinationen*. Zurich: Edition Olms.

Sapolsky, R. M. (1996). Stress, glucocorticoids, and damage to the nervous system: the current state of confusion. *Stress*, *1*(1), 1–19.

Scarr, S., and McCartney, K. (1983). How people make their own environments: a theory of genotype → environment effects. *Child Development*, *54*(2), 424–435.

Schaie, K. W. (1994). The course of adult intellectual development. *American Psychologist*, *49*(4), 304–313.

Schmitt, D. P., Realo, A., Voracek, M., and Allik, J. (2008). Why can't a man be more like a woman? Sex differences in big five personality traits across 55 cultures. *Journal of Personality and Social Psychology*, *94*(1), 168–182.

Schneider, W., Gruber, H., Gold, A., and Opwis, K. (1993). Chess expertise and memory for chess positions in children and adults. *Journal of Experimental Child Psychology*, *56*(3), 328–349.

Scholz, M., Niesch, H., Steffen, O., Ernst, B., Loeffler, M., Witruk, E., et al. (2008). Impact of chess training on mathematics performance and concentration ability of children with learning disabilities. *International Journal of Special Education*, *23*(3), 138–148.

Schultetus, R., and Charness, N. (1999). Recall or evaluation of chess positions revisited: the relationship between memory and evaluation in chess skill. *American Journal of Psychology, 112*(4), 555–569.

Schwarz, A. M., Schachinger, H., Adler, R. H., and Goetz, S. M. (2003). Hopelessness is associated with decreased heart rate variability during championship chess games. *Psychosomatic Medicine, 65*(4), 658–661.

Scurrah, M. J., and Wagner, D. A. (1970). Cognitive model of problem-solving in chess. *Science, 169*, 209–211.

Selye, H. (1975). Stress and distress. *Comprehensive Therapy, 1*(8), 9–13.

Shannon, C. E. (1950). Programming a computer for playing chess. *Philosophical Magazine, 41*(314), 256–275.

Shanteau, J. (1992). Competence in experts: the role of task characteristics. *Organizational Behavior and Human Decision Processes, 53*(2), 252–266.

Shanteau, J. (2015). Why task domains (still) matter for understanding expertise. *Journal of Applied Research in Memory and Cognition, 4*(3), 169–175.

Sherman, R. A., Nave, C. S., and Funder, D. C. (2010). Situational similarity and personality predict behavioral consistency. *Journal of Personality and Social Psychology, 99*(2), 330–343.

Sigirtmac, A. D. (2012). Does chess training affect conceptual development of six-year-old children in Turkey? *Early Child Development and Care, 182*(6), 797–806.

Sigman, M., Etchemendy, P., Fernández-Slezak, D., and Cecchi, G. A. (2010). Response time distributions in rapid chess: a large-scale decision making experiment. *Frontiers in Neuroscience, 4*, article 60.

Silva-Junior, L. R., Cesar, F. H. G., Rocha, F. T., and Thomaz, C. E. (2018). A combined eye-tracking and EEG analysis on chess moves. *IEEE Latin American Transactions, 16*(5), 1288–1297.

Silver, D., Hubert, T., Schrittwieser, J., Antonoglou, I., Lai, M., Guez, A., et al. (2018). A general reinforcement learning algorithm that masters chess, shogi, and Go through self-play. *Science, 362*, 1140–1144.

Simon, H. A., and Barenfeld, M. (1969). Information-processing analysis of perceptual processes in problem solving. *Psychological Review, 76*(5), 473–483.

Simon, H. A., and Chase, W. G. (1973). Skill in chess. *American Scientist, 61*(4), 394–403.

Simon, H. A., and Gilmartin, K. (1973). A simulation of memory for chess positions. *Cognitive Psychology, 5*(1), 29–46.

Simonton, D. K. (1997). Creative productivity: a predictive and explanatory model of career trajectories and landmarks. *Psychological Review, 104*(1), 66–89.

Simonton, D. K. (1999). Significant samples: the psychological study of eminent individuals. *Psychological Methods, 4*(4), 425–451.

Simonton, D. K. (2000). Creative development as acquired expertise: theoretical issues and an empirical test. *Developmental Review, 20*(2), 283–318.

Simonton, D. K. (2006). Historiometric methods. In K. A. Ericsson, N. Charness, P. J. Feltovich, and R. R. Hoffman (eds.), *The Cambridge Handbook of Expertise and Expert Performance*. New York: Cambridge University Press, 319–335.

Simonton, D. K. (2014a). Creative performance, expertise acquisition, individual differences, and developmental antecedents: an integrative research agenda. *Intelligence*, *45*, 66–73.

Simonton, D. K. (2014b). Addressing the recommended research agenda instead of repeating prior arguments. *Intelligence*, *45*, 120–121.

Skanes, G. R., Sullivan, A. M., Rowe, E. J., and Shannon, E. (1974). Intelligence and transfer: aptitude by treatment interactions. *Journal of Educational Psychology*, *66*(4), 563–568.

Smillie, L. D. (2008). What is reinforcement sensitivity? Neuroscience paradigms for approach–avoidance process theories of personality. *European Journal of Personality*, *22*(5), 359–384.

Snow, R. E., and Lohman, D. F. (1984). Toward a theory of cognitive aptitude for learning from instruction. *Journal of Educational Psychology*, *76*(3), 347–376.

Snow, R. E., and Swanson, J. (1992). Instructional psychology: aptitude, adaptation, and assessment. *Annual Review of Psychology*, *43*, 583–626.

Soldz, S., and Vaillant, G. E. (1999). The big five personality traits and the life course: a 45-year longitudinal study. *Journal of Research in Personality*, *33*(2), 208–232.

Sonas, J. (2002). The Sonas rating formula: better than Elo? ChessBase Magazine Online, 22 October, www.chessbase.com/newsdetail.asp?newsid=562.

Spearman, C. (1904). 'General intelligence,' objectively determined and measured. *American Journal of Psychology*, *15*(2), 201–293.

Spearman, C. (1927). *The Abilities of Man: Their Nature and Measurement*. New York: Macmillan.

Specht, J., Egloff, B., and Schmukle, S. C. (2011). Stability and change of personality across the life course: the impact of age and major life events on mean-level and rank-order stability of the big five. *Journal of Personality and Social Psychology*, *101*(3), 862–882.

Spelke, E. S. (2005). Sex differences in intrinsic aptitude for mathematics and science? *American Psychologist*, *60*(9), 950–958.

Stafford, T. (2018). Female chess players outperform expectations when playing men. *Psychological Science*, *29*(3), 429–436.

Stanovich, K. E., and West, R. F. (2000). Individual differences in reasoning: implications for the rationality debate? *Behavioral and Brain Sciences*, *23*(5), 645–726.

Steele, C. M. (1997). A threat in the air: how stereotypes shape intellectual identity and performance. *American Psychologist*, *52*(6), 613–629.

Stepien, P., Klonowski, W., and Suvorov, N. (2015). Nonlinear analysis of EEG in chess players. *EPJ NonLinear Biomedical Physics*, *3*, article 1.

Stern, W. (1921). *Die psychologische Methoden der Intelligenzprüfung*. Leipzig: Barth.

Sternberg, R. J. (1999). The theory of successful intelligence. *Review of General Psychology*, *3*(4), 292–316.

Sternberg, R. J., and Kaufman, S. B. (eds.) (2011). *The Cambridge Handbook of Intelligence*. New York: Cambridge University Press.

Stoet, G., and Geary, D. C. (2012). Can stereotype threat explain the gender gap in mathematics performance and achievement? *Review of General Psychology*, *16* (1), 93–102.

Stoet, G., and Geary, D. C. (2015). Sex differences in academic achievement are not related to political, economic, or social equality. *Intelligence*, *48*, 137–151.

Storey, K. (2000). Teaching beginning chess skills to students with disabilities. *Preventing School Failure*, *44*(2), 45–49.

Su, R., Rounds, J., and Armstrong, P. I. (2009). Men and things, women and people: a meta-analysis of sex differences in interests. *Psychological Bulletin*, *135*(6), 859–884.

Swider, B. W., and Zimmerman, R. D. (2010). Born to burnout: a meta-analytic path model of personality, job burnout, and work outcomes. *Journal of Vocational Behavior*, *76*(3), 487–506.

Sykes, E. D. A., Bell, J. F., and Vidal Rodeiro, C. (2016). *Birthdate Effects: A Review of the Literature from 1990-on*. Cambridge: Local Examinations Syndicate, University of Cambridge.

Tanaka, W. J., and Taylor, M. (1991). Object categories and expertise: is the basic level in the eye of the beholder? *Cognitive Psychology*, *23*(3), 457–482.

Thomas, R. P., and Lawrence, A. (2018). Assessment of expert performance compared across professional domains. *Journal of Applied Research in Memory and Cognition*, *7*(2), 167–176.

Thompson, M. (2003). Does the playing of chess lead to improved scholastic achievement? *Issues in Educational Research*, *13*(2), 13–26.

Thorndike, E. L. (1908). The effect of practice in the case of purely intellectual function. *American Journal of Psychology*, *19*(3), 374–384.

Thurstone, L. L. (1938). *Primary Mental Abilities*. Chicago: University of Chicago Press.

Tikhomirov, O. K., and Poznyanskaya, E. D. (1966). An investigation of visual search as a means of analyzing heuristics. *Soviet Psychology*, *5*(2), 3–15.

Tikhomirov, O. K., and Vinogradov, Y. E. (1970). Emotions in the function of heuristics. *Soviet Psychology*, *8*(3/4), 198–223.

Tooby, J., and Cosmides, L. (1990). On the universality of human nature and the uniqueness of the individual: the role of genetics and adaptation. *Journal of Personality*, *58*(1), 17–68.

Tooby, J., and Cosmides, L. (2005). Conceptual foundations of evolutionary psychology. In D. M. Buss (ed.), *The Handbook of Evolutionary Psychology*. Hoboken, NJ: Wiley, 5–67.

Tooby, J., Cosmides, L., and Barrett, H. C. (2003). The second law of thermodynamics is the first law of psychology: evolutionary developmental psychology and the theory of tandem, coordinated inheritances: comment on Lickliter and Honeycutt (2003). *Psychological Bulletin, 129*(6), 858–865.

Trahan, L. H., Stuebing, K. K., Fletcher, J. M., and Hiscock, M. (2014). The Flynn effect: a meta-analysis. *Psychological Bulletin, 140*(5), 1332–1360.

Trinchero, R., and Sala, G. (2016). Chess training and mathematical problem solving: the role of teaching heuristics in transfer of learning. *Eurasia Journal of Mathematics, Science and Technology Education, 12*(3), 655–668.

Troubat, N., Fargeas-Gluck, M. A., Tulppo, M., and Dugué, B. (2009). The stress of chess players as a model to study the effects of psychological stimuli on physiological responses: an example of substrate oxidation and heart rate variability in man. *European Journal of Applied Physiology, 105*(3), 343–349.

Tucker-Drob, E. M., and Briley, D. A. (2014). Continuity of genetic and environmental influences on cognition across the life span: a meta-analysis of longitudinal twin and adoption studies. *Psychological Bulletin, 140*(4), 949–979.

Tucker-Drob, E. M., Briley, D. A., and Harden, K. P. (2013). Genetic and environmental influences on cognition across development and context. *Current Directions in Psychological Science, 22*(5), 349–355.

Tukey, J. (1949). Comparing individual means in the analysis of variance. *Biometrics, 5*(2), 99–114.

Tupes, E. C., and Christal, R. E. (1992). Recurrent personality factors based on trait ratings. *Journal of Personality, 60*(2), 225–251.

Turkheimer, E. (1991). Individual and group differences in adoption studies of IQ. *Psychological Bulletin, 110*(3), 392–405.

Turkheimer, E. (1998). Heritability and biological explanation. *Psychological Review, 105*(4), 782–791.

Turkheimer, E. (2000). Three laws of behavior genetics and what they mean. *Current Directions in Psychological Science, 9*(5), 160–164.

Turkheimer, E., and Waldron, M. (2000). Nonshared environment: a theoretical, methodological, and quantitative review. *Psychological Bulletin, 126*(1), 78–108.

Uher, J. (2008). Comparative personality research: methodological approaches. *European Journal of Personality, 22*(5), 427–455.

Ullén, F., Hambrick, D. Z., and Mosing, M. A. (2016). Rethinking expertise: a multifactorial gene–environment interaction model of expert performance. *Psychological Bulletin, 142*(4), 427–446.

Unterrainer, J. M., Kaller, C. P., Halsband, U., and Rahm, B. (2006). Planning abilities and chess: a comparison of chess and non-chess players on the Tower of London task. *British Journal of Psychology, 97*(3), 299–311.

Unterrainer, J. M., Kaller, C. P., Leonhart, R., and Rahm, B. (2011). Revising superior planning performance in chess players: the impact of time restriction and motivation aspects. *American Journal of Psychology, 124*(2), 213–225.

Urbina, S. (2011). Tests of intelligence. In R. J. Sternberg and S. B. Kaufman (eds.), *The Cambridge Handbook of Intelligence*. New York: Cambridge University Press, 20–38.

Vaci, N., and Bilalić, M. (2017). Chess databases as a research vehicle in psychology: modeling large data. *Behavioral Research*, *49*(4), 1227–1240.

Vaci, N., Gula, B., and Bilalić, M. (2015). Is age really cruel to experts? Compensatory effects of activity. *Psychology and Aging*, *30*(4), 740–754.

Van der Maas, H. L. J., Dolan, C. V., Grasman, R. P. P. P., Wicherts, J. M., Huizenga, H. M., and Raijmakers, M. E. J. (2006). A dynamical model of general intelligence: the positive manifold of intelligence by mutualism. *Psychological Review*, *113*(4), 842–861.

Van der Maas, H. L. J., and Wagenmakers, E. J. (2005). A psychometric analysis of chess expertise. *American Journal of Psychology*, *118*(1), 29–60.

Van der Stigchel, S., and Hollingworth, A. (2018). Visuospatial working memory as a fundamental component of the eye movement system. *Current Directions in Psychological Science*, *27*(2), 136–143.

Van Harreveld, F., Wagenmakers, E. J., and van der Maas, H. L. J. (2007). The effects of time pressure on chess skill: an investigation into fast and slow processes underlying expert performance. *Psychological Research*, *71*(5), 591–597.

Vasyukova, E. E. (2012). The nature of chess expertise: knowledge or search? *Psychology in Russia: State of the Art*, *5*, 511–528.

Velarde-Lombraña, J. (1993). Huarte de San Juan, patrono de psicología. *Psicothema*, *5*(2), 451–458.

Vernon, P. E. (1950). *The Structure of Human Abilities*. New York: Wiley.

Vicente, K. J., and de Groot, A. D. (1990). The memory recall paradigm: straightening out the historical record. *American Psychologist*, *45*(2), 285–287.

Volke, H. J., Dettmar, P., Richter, P., Rudolf, M., and Buhss, U. (2002). On-coupling and off-coupling of neocortical areas in chess experts and novices as revealed by evoked EEG coherence measures and factor-based topological analysis: a pilot study. *Journal of Psychophysiology*, *16*(1), 23–36.

Volkow, N. D., Rosen, B., and Farde, L. (1997). Imaging the living human brain: magnetic resonance imaging and positron emission tomography. *Proceedings of the National Academy of Sciences*, *94*(7), 2787–2788.

Vollstädt-Klein, S., Grimm, O., Kirsch, P., and Bilalić, M. (2010). Personality of elite male and female chess players and its relation to chess skill. *Learning and Individual Differences*, *20*(3), 517–521.

Voyer, D., Rousseau, M., Rajotte, T., Freiman, V., and Cabot-Thibault, J. (2018). Le jeu d'échecs dans les classes du primaire: un moyen ciblé pour développer les habiletés des garçons en résolution de problèmes. *Formation et Profession*, *26*(3), 94–108.

Voyer, D., and Voyer, S. D. (2014). Gender differences in scholastic achievement: a meta-analysis. *Psychological Bulletin*, *140*(4), 1174–1204.

Vukasovic, T., and Bratko, D. (2015). Heritability of personality: a meta-analysis of behavior genetic studies. *Psychological Bulletin, 141*(4), 769–785.

Wagner, D. A., and Scurrah, M. J. (1971). Some characteristics of human problem-solving in chess. *Cognitive Psychology, 1*(4), 454–478.

Wai, J. (2013). Investigating America's elite: cognitive ability, education, and sex differences. *Intelligence, 41*(4), 203–211.

Wai, J. (2014a). Experts are born, then made: combining prospective and retrospective longitudinal data shows that cognitive ability matters. *Intelligence, 45*, 74–80.

Wai, J. (2014b). What does it mean to be an expert? *Intelligence, 45*, 122–123.

Wai, J., Cacchio, M., Putallaz, M., and Makel, M. C. (2010). Sex differences in the right tail of cognitive abilities: a 30 year examination. *Intelligence, 38*(4), 412–423.

Wan, X., Nakatani, H., Ueno, K., Asamizuya, T., Cheng, K., and Tanaka, K. (2011). The neural basis of intuitive best next-move generation in board game experts. *Science, 331*, 341–346.

Wang, M. T., Eccles, J. S., and Kenny, S. (2013). Not lack of ability but more choice: individual and gender differences in choice of careers in science, technology, engineering, and mathematics. *Psychological Science, 24*(5), 770–775.

Waters, A. J., and Gobet, F. (2008). Mental imagery and chunks: empirical and computational findings. *Memory and Cognition, 36*(3), 505–517.

Waters, A. J., Gobet, F., and Leyden, G. (2002). Visuospatial abilities of chess players. *British Journal of Psychology, 93*(3), 557–565.

Weiergräber, M., Papazoglou, A., Broich, K., and Müller, R. (2016). Sampling rate, signal bandwidth and related pitfalls in EEG analysis. *Journal of Neuroscience Methods, 268*, 53–55.

Wechsler, D. (2003). *Wechsler Intelligence Scale for Children*, 4th edn. San Antonio, TX: Pearson.

Wechsler, D. (2008). *Wechsler Adult Intelligence Scale*, 4th edn. San Antonio, TX: Pearson.

Wonderlic, E. F., and Wonderlic, C. F. (1992). *Wonderlic Personnel Test User's Manual*. Libertyville, IL; Wonderlic Personnel Test.

Wood, W., and Eagly, A. H. (2002). A cross-cultural analysis of the behavior of women and men: implications for the origins of sex differences. *Psychological Bulletin, 128*(5), 699–727.

Woodcock, R. W., McGrew, K. S., and Mather, N. (2001). *Woodcock-Johnson® III Test*. Itasca, IL: Riverside Publishing.

Wright, M. J., Gobet, F., Chassy, P., and Ramchandani, P. N. (2013). ERP to chess stimuli reveal expert–novice differences in the amplitudes of N2 and P3 components. *Psychophysiology, 50*(10), 1023–1033.

Yamagata, S., Suzuki, A., Ando, J., Ono, Y., Kijima, N., Yoshimura, K., et al. (2006). Is the genetic structure of human personality universal? A cross-cultural twin study from North America, Europe and Asia. *Journal of Personality and Social Psychology, 90*(6), 987–998.

Yap, K. O. (2006). *Chess for Success Evaluation*. Portland, OR: Northwest Regional Educational Laboratory.

Yoskowitz, J. (1991). Chess versus quasi-chess: the role of knowledge of legal rules. *American Journal of Psychology*, *104*(3), 355–366.

Yuan, K. H., and Bentler, P. (2004). On chi-square difference and z tests in mean and covariance structure analysis when the base model is misspecified. *Educational and Psychological Measurement*, *64*(5), 737–757.

Zuckerman, M. (1991). *Psychobiology of Personality*. New York: Cambridge University Press.

Zuckerman, M. (2005). *Psychobiology of Personality*, 2nd edn. New York: Cambridge University Press.

Zuckerman, M., Kuhlman, M. D., Joireman, J., Teta, P., and Kraft, M. (1993). A comparison of three structural models for personality: the big three, the big five, and the alternative five. *Journal of Personality and Social Psychology*, *65*(4), 757–768.

# INDEX

Lightning Source UK Ltd.
Milton Keynes UK
UKHW020828101222
413700UK00031B/640

9 781108 469456